ENERGY OPTIMIZATION PROTOCOL DESIGN FOR SENSOR NETWORKS IN IOT DOMAINS

This book provides an essential overview of Internet of things (IoT), energy-efficient topology control protocols, motivation, and challenges for topology control for wireless sensor networks, and the scope of the research in the domain of IoT. Further, it discusses the different design issues of topology control and energy models for IoT applications, and different types of simulators with their advantages and disadvantages. It also discusses extensive simulation results and comparative analysis for various algorithms. The key point of this book is to present a solution to minimize energy and extend the lifetime of IoT networks using optimization methods to improve the performance.

Features:

- Describes various facets necessary for energy optimization in IoT domain.
- Covers all aspects to achieve energy optimization using the latest technologies and algorithms, in wireless sensor networks.
- Presents various IoT and topology control methods and protocols, various network models, and model simulations using MATLAB®.
- Reviews methods and results of optimization with Simulation Hardware architecture leading to prolonged life of IoT networks.
- First time introduces bio-inspired algorithms in the IoT domain for performance optimization.

This book aims at graduate students, researchers in information technology, computer science and engineering, electronics and communication engineering.

ENERGY OPTIMIZATION PROTOCOL DESIGN FOR SENSOR NETWORKS IN IOT DOMAINS

Sanjeev J. Wagh, Manisha Sunil Bhende, and Anuradha D. Thakare

CRC Press
Taylor & Francis Group
Boca Raton London New York

CRC Press is an imprint of the
Taylor & Francis Group, an **informa** business

MATLAB® is a trademark of The MathWorks, Inc. and is used with permission. The MathWorks does not warrant the accuracy of the text or exercises in this book. This book's use or discussion of MATLAB® software or related products does not constitute endorsement or sponsorship by The MathWorks of a particular pedagogical approach or particular use of the MATLAB® software.

First edition published 2023
by CRC Press
6000 Broken Sound Parkway NW, Suite 300, Boca Raton, FL 33487-2742

and by CRC Press
4 Park Square, Milton Park, Abingdon, Oxon, OX14 4RN

CRC Press is an imprint of Taylor & Francis Group, LLC

© 2023 Sanjeev J. Wagh, Manisha Sunil Bhende and Anuradha D. Thakare

Reasonable efforts have been made to publish reliable data and information, but the author and publisher cannot assume responsibility for the validity of all materials or the consequences of their use. The authors and publishers have attempted to trace the copyright holders of all material reproduced in this publication and apologize to copyright holders if permission to publish in this form has not been obtained. If any copyright material has not been acknowledged please write and let us know so we may rectify in any future reprint.

Except as permitted under U.S. Copyright Law, no part of this book may be reprinted, reproduced, transmitted, or utilized in any form by any electronic, mechanical, or other means, now known or hereafter invented, including photocopying, microfilming, and recording, or in any information storage or retrieval system, without written permission from the publishers.

For permission to photocopy or use material electronically from this work, access www.copyright.com or contact the Copyright Clearance Center, Inc. (CCC), 222 Rosewood Drive, Danvers, MA 01923, 978-750-8400. For works that are not available on CCC please contact mpkbookspermissions@tandf.co.uk

Trademark notice: Product or corporate names may be trademarks or registered trademarks and are used only for identification and explanation without intent to infringe.

Library of Congress Cataloging-in-Publication Data
A catalog record for this title has been requested

ISBN: 978-1-032-30559-2 (hbk)
ISBN: 978-1-032-31611-6 (pbk)
ISBN: 978-1-003-31054-9 (ebk)

DOI: 10.1201/9781003310549

Typeset in Times
by MPS Limited, Dehradun

Contents

Preface xi
Author's Biography xiii
Abbreviations xv

Chapter 1 Introduction and Background Study 1
    1.1 IoT and WSN 1
        1.1.1 Overview of WSN 1
        1.1.2 How Does WSN Works? 2
        1.1.3 Security Issues in WSN 7
    1.2 IoT and Sensor Network Applications 8
        1.2.1 Wide Space Applications 8
        1.2.2 Small Space Application 9
    1.3 OSI and IoT Layer Stack 9
        1.3.1 Physical or Sensor Layer 10
        1.3.2 Processing and Control Layer 10
        1.3.3 Hardware Interface Layer 11
        1.3.4 RF Layer 11
        1.3.5 Session/Message Layer 11
        1.3.6 User Experience Layer 11
        1.3.7 Application Layer 11
    1.4 Protocols in WSN and IoT 11
        1.4.1 Routing Protocol for Low-Power and Lossy Networks 12
        1.4.2 Cognitive RPL 12
        1.4.3 Lightweight On-Demand AD hoc Distance Vector Routing – Next Generation (LOAD$_{\text{NG}}$) 12
        1.4.4 Collection Tree Protocol 13
        1.4.5 Channel-Aware Routing Protocol 13
        1.4.6 E-CARP 14
    1.5 Energy Consumption and Network Topology 14
    1.6 Challenges for Energy Consumption in IoT Networks 14
        1.6.1 Energy Consumption 15
        1.6.2 Combination of IoT with Subsystems 15
        1.6.3 User Privacy 15
        1.6.4 Safety Challenge 15
        1.6.5 IoT Standards 16
        1.6.6 Architecture Design 16
    1.7 Summary 16
    References 17

| Chapter 2 | IoT and Topology Control: Methods and Protocol | 19 |

2.1 Sensor Network Topologies 19
    2.1.1 Star Network (Single Point-to-Multipoint) 19
    2.1.2 Mesh Network Topology 20
    2.1.3 Hybrid-Star-Mesh Network Topology 20
2.2 IoT and Topology Control Methods 21
    2.2.1 Powder Adjustment Approach 23
    2.2.2 Powder Mode Approach 26
    2.2.3 Clustering Approach 31
    2.2.4 Hybrid Approach 36
2.3 Comparative Analysis: Topology Control Methods 41
    2.3.1 Evaluations Based on the Network Lifetime Definitions 44
    2.3.2 Evaluations Based on the Network Lifetime Definitions 47
    2.3.3 Evaluations Based on the Network Lifetime Definitions 49
    2.3.4 Evaluations Based on the Network Lifetime Definitions 51
2.4 IoT and Topology Control Protocols 53
    2.4.1 Link Efficiency-Based Topology Control 53
    2.4.2 Improved Reliable and Energy Efficient Topology Control 54
    2.4.3 Cellular Automata-Based Topology Control 54
    2.4.4 Heterogeneous Topology Control Algorithm (HTC) 54
2.5 IoT and Routing Protocols 55
    2.5.1 Routing Protocol for Low-Power and Lossy Networks (RPL) 55
    2.5.2 Cognitive RPL (CORP) 59
    2.5.3 Channel-Aware Routing Protocol (CARP) 60
2.6 Future Research Direction: Context-Aware Routing in IoT Networks 60
    2.6.1 Routing in IoT 60
    2.6.2 Need of Context-Awareness in IoT Routing 61
    2.6.3 Context Needed for Routing 61
2.7 Summary 64
References 64

| Chapter 3 | Design Issues, Models, and Simulation Platforms | 69 |

3.1 Topology Control Design Issues 69
    3.1.1 Taxonomy of Topology Issues 70
    3.1.2 Topology Awareness Problem 71
    3.1.3 Topology Control Problem 73
3.2 Network Models 76

		3.2.1	Homogeneous Model	76
		3.2.2	Wireless Propagation Model	77
		3.2.3	Model of Long-Distance Path	77
		3.2.4	Hop Model	78
		3.2.5	Energy Model	78
	3.3	Simulation Platforms		79
		3.3.1	OMNeT++	79
		3.3.2	NS2	80
	3.4	Simulation Using MATLAB for IoT Domain		81
		3.4.1	The MATLAB System	81
		3.4.2	MATLAB for IoT Domain	82
	3.5	Future Research Direction: Heterogeneity of Network Technologies		82
		3.5.1	Sensing Layer	83
		3.5.2	Network Layer	84
		3.5.3	Cloud Computing	84
		3.5.4	Application Layer	84
		3.5.5	Smart Industrial	85
		3.5.6	Smart Agricultural	86
		3.5.7	Smart Home	86
		3.5.8	Intelligent Transportation System	87
		3.5.9	Smart Healthcare	87
	3.6	Summary		88
	References			88
Chapter 4	Link Efficiency-Based Topology Control Algorithm for IoT Domain Application			93
	4.1	Introduction		93
		4.1.1	Received Signal Strength Indicator	93
		4.1.2	Limitation of RSSI	94
	4.2	Network Model		94
		4.2.1	Definitions	95
		4.2.2	Assumptions	96
	4.3	Improved Link Efficiency-Based Topology Control Algorithm		96
		4.3.1	Proposed Algorithm: LEBTC	97
		4.3.2	Mathematical Model	98
		4.3.3	Flow Diagram	99
	4.4	Implementations		99
		4.4.1	RNG-Relative Neighborhood Graph	99
		4.4.2	GG – Gabriel Graph	99
		4.4.3	FETC and FETCD	99
	4.5	Future Research Direction: Gateway Placement and Energy-Efficient Scheduling in IoT		110

	4.5.1	Overview	110
	4.5.2	Placement of Gateways	115
	4.5.3	Task Model	116
	4.5.4	Energy Consumption Model	116
	4.5.5	Energy-Efficient Scheduling Algorithms	116
4.6	Summary		118
References			119

Chapter 5 Energy-Efficient Topology Control Algorithms for IoT Domain Applications ... 121

5.1	Introduction	121
	5.1.1 Connected Dominating Set	121
	5.1.2 Clustering Mechanisms	125
5.2	Network Model	125
5.3	Energy-Efficient Algorithm Based on Connected Dominating Set	126
	5.3.1 Proposed Algorithm: iPOLY	127
	5.3.2 Mathematical Model	127
	5.3.3 Flow Diagrams	129
5.4	Implementations: POLY and iPOLY	129
5.5	Future Research Direction: IoT Reliability	130
	5.5.1 Device Reliability	130
	5.5.2 Network Reliability	133
	5.5.3 System Reliability	133
	5.5.4 Anomaly Detection	134
5.6	Summary	134
References		134

Chapter 6 Cellular Automata-Based Topology Control Algorithms for IoT Domain Applications ... 137

6.1	Introduction	137
	6.1.1 Cellular Automata for Sensor Networks	139
	6.1.2 Sensor Network Clustering	140
6.2	Cellular Automata-Based Topology Control Algorithms	140
	6.2.1 Cellular Automata Weighted Margoles Neighborhood	140
	6.2.2 Cellular Automata Weighted Moor Neighborhood	141
	6.2.3 Cyclic Cellular Automata	143
6.3	Future Research Direction: Cellular Automata for IoT Application	152
6.4	Summary	153
References		153

Contents

Chapter 7 Performance Optimization in IoT Networks 155
- 7.1 IoT Network Issues 155
  - 7.1.1 Fault Tolerance 155
  - 7.1.2 Security Enforcement 156
  - 7.1.3 Handling Heterogeneity 157
  - 7.1.4 Self-Configuration 158
  - 7.1.5 Unintended Interference 158
  - 7.1.6 Network Visibility 160
  - 7.1.7 Restricted Access 160
- 7.2 Optimization Issues in IoT Networks 160
  - 7.2.1 Data Aggregation 162
  - 7.2.2 Routings in IoT Networks 163
  - 7.2.3 Optimal Coverage 164
  - 7.2.4 Sensor Localization 165
- 7.3 Optimization Levels in IoT 166
  - 7.3.1 Device Level Optimizationn 167
  - 7.3.2 Network Level Optimization 168
  - 7.3.3 Application Level Optimization 170
- 7.4 Solutions for IoT Network Optimization 171
  - 7.4.1 Network Routing 172
  - 7.4.2 Energy Conservation 172
  - 7.4.3 Congestion Control 174
  - 7.4.4 Heterogeneity 174
  - 7.4.5 Scalability 175
  - 7.4.6 Network Reliability 176
  - 7.4.7 Quality of Service 177
- 7.5 Summary 178
- References 179

Chapter 8 Bio-Inspired Computing and IoT Networks 181
- 8.1 Bio-Inspired Approach 181
  - 8.1.1 Bio-Inspired Computing 181
  - 8.1.2 Bio-Inspired System 181
  - 8.1.3 Bio-Inspired Engineering 183
- 8.2 Motivation for Bio-Inspired Computing 184
  - 8.2.1 Self-Organization 184
  - 8.2.2 Self-Adaptation 185
  - 8.2.3 Self-Healing Ability 186
- 8.3 Bio-Inspired Computing Approaches for Optimizations 188
  - 8.3.1 Evolutionary Algorithms (EAs) 189
  - 8.3.2 Artificial Neural Networks (ANNs) 193
  - 8.3.3 Swarm Intelligence (SI) 196
  - 8.3.4 Firefly Algorithm (FA) 198
  - 8.3.5 Artificial Immune System (AIS) 199
  - 8.3.6 Epidemic Spreading (ES) 201

| | | 8.4 | Summary .... 203 |
| | | References .... 204 |

Chapter 9 Blockchain and IoT Optimization 205

- 9.1 Blockchain Technology and IoT 205
  - 9.1.1 Introduction to Blockchain 205
  - 9.1.2 Blockchain Terminology 205
  - 9.1.3 Blockchain Mechanism 206
  - 9.1.4 Distributed P2P Networking 208
- 9.2 Blockchain Support for IoT Applications 209
  - 9.2.1 Securing IoT Networks 209
  - 9.2.2 Manufacturing Maintenance Support 210
  - 9.2.3 Transparency in Supply Chain 212
  - 9.2.4 In-Car Payment Model 212
  - 9.2.5 Vehicle Insurance Model 213
  - 9.2.6 Identity Authentication Using Self-Sovereign Identity (SSI) 213
- 9.3 Blockchain with IoT Networks Characteristics 214
  - 9.3.1 Security 215
  - 9.3.2 Scalability 215
  - 9.3.3 Immutability and Auditing 217
  - 9.3.4 Effectiveness and Efficiency 217
  - 9.3.5 Traceability and Interoperability 219
  - 9.3.6 Quality of Service 220
- 9.4 Energy Optimization and Blockchain Mechanism 220
  - 9.4.1 Optimization Process 221
  - 9.4.2 Resource Management Using Blockchain 221
- 9.5 Energy Optimization in Blockchain-Enabled IoT Networks 222
- 9.6 Summary 223

References 223

Index 225

Preface

Internet of things (IoT) and wireless sensor networks (WSN) are the fastest-growing technologies for deploying general applications ranging from smart home, retail, supply chain, smart city, industrial Internet, smart farming, and connected cars to healthcare. IoT has now become a significant empowering technology that extends various technology dominions from data sensing and processing to networking and data analytics. This book addresses components of IoT, the importance of topology control in IoT and WSN, energy-efficient topology control protocols to improve the performance of IoT, and a framework for performance optimization in the IoT. The development of IoT applications and services includes various components such as sensors, actuators, gateways, cloud, etc. Sensors are a very important component of the IoT system.

In this book, the authors provide an essential overview of IoT, energy-efficient topology control protocols, motivation, and challenges for topology control for wireless sensor networks, the scope of the research in the domain of IoT is also presented. The aims and objectives of the research are elucidated. Scientific contributions of the research work in the IoT domain are explored. Further authors discuss the different design issues of topology control and energy models for IoT applications. Different types of simulators with their advantages and disadvantages are discussed. This book provides a discussion on the results for contribution carried out with extensive simulation results and comparative analysis for various algorithms.

There are still many aspects of topology control for sensor networks that are not considered in this book because some of them are out of scope or due to limitations of time and resources. There are plenty of rooms that can be explored and added on top of our proposed algorithms. Among the issues that are evaluated in the center of attention of this book, there are still several aspects for further research.

Today, most system designers imagine things to create or design every possible system based on the IoT. The components in the IoT system have embedded computing technologies that support internet facilities to support many application domains. WSN and IoT have become the prominent field of study for researchers because of their readily available and applicable resources and extensive application over the various fields.

IoT is a network of internet-enabled sensor devices and different heterogeneous wireless technologies that are used for data communication. The analyst firm Gartner says that by 2021 there will be over 26 billion connected devices and a lot of connections. IoT involves billions of connected sensors communicating through the wireless network, energy efficiency and fairness is the biggest concern. It raises a need to develop energy-efficient techniques or approach that reduces contention and provide energy-efficient communication. The IoT is a giant network of connected things. Many of the objects in the surrounding environment are on the network in different forms. And enormous amounts of data are generated which have to be stored, processed, and presented in a seamless, efficient, and easily interpretable form. In a real-world scenario, IoT collects data by various sensor devices and disseminates sensed data through gateway devices to the IoT Cloud (or the Internet) either using single or multi-hop communication.

The main function of any smart product application is how they communicate with each other. For this purpose, many topology and routing protocols have been devised to minimize the differences over the head of the system like energy, cost, distance, etc. Routing for different wireless sensor applications based on parameters is a smooth and dependent solution resulting in low costs information for the network package, as well as for the node itself. This project introduces the kind of protocol for using a different neighbor's node and different parameters which are associated with the system like distance, energy, cost, etc. The key point of this book is to present a solution to minimize energy and extend the lifetime of IoT networks and present optimization methods to improve the performance.

The literature in this book explores implementations of new techniques and algorithms that will support achieving significant enhancement in the existing IoT-based techniques in the domain. This book will be interpreted as a place to begin and a useful comparative reference for those inquisitive about the continuously evolving subject of the IoT.

Author's Biography

Sanjeev J. Wagh, works as **Professor and Head** in the Department of Information Technology at **Govt. College of Engineering, Karad**. He completed his BE (1996), ME (2000), and Ph.D. (2009) in Computer Science and Engineering from Govt. College of Engineering, Pune and Nanded. He was a full-time **Post Doctorate fellow** at the Center for TeleInfrastructure, **Aalborg University, Denmark** during 2013–14. He has also completed an MBA (IT) from NIBM (2015), Chennai. He has a total of 24 years of experience in academics and research. His research interest areas are natural science computing, internet technologies, and wireless sensor networks, data sciences and analytics. He has more than **100 research papers** to his credit, published in international/national journals and conferences. Four research scholars completed their Ph.D. under his supervision from Pune University. Currently, three research scholars are pursuing Ph.D.'s under his supervision in various Indian Universities. He is a fellow member of ISTE, IETE, and a member of IEEE, ACM, and CSI. He is co-editor for *International Journals in Engineering & Technology*. He has visited Denmark (Aalborg University, Aalborg and Copenhagen), Sweden (Gothenburg University, Gothenburg), Germany (Hamburg University, Hamburg), Norway (University of Oslo), France (the University of London Institute in Paris), China (Shanghai Technology Innovation Center Shanghai, delegation visit), Thailand (Kasetsart University, Bangkok), Mauritius (University of Technology, Port Louis) for academics and research. He authored the book *"Fundamentals of Data Science"* published by CRC Press and edited the book *"Applied Machine Learning for Smart Data Analysis"* published by CRC Press and *"Handbook of Research on Applied Intelligence for Health and Clinical Informatics"* published by IGI Global, USA.

Dr. Manisha S. Bhende, works as a Professor at Marathwada Mitra Mandals Institute of Technology, Pune, India. She has completed a BE (1998), ME (2007), and Ph.D. (2017) in Computer Engineering from the University of Pune and a bachelor's degree from Government College of Engineering, Amravati, India. Her research interests are IoT and wireless networks, network security, cloud computing, data science and machine learning, data analytics, etc. She has 49 research papers/book chapters in international and national conferences and journals. She delivered an expert talk on various domains such as wireless communication, wireless sensor networks, data science, cyber security, IoT, embedded and real-time operating systems, IPR and innovation, etc. She has published 4 patents and received 5 copyrights on her credit. She is a reviewer/examiner for a Ph.D. thesis and ME dissertations for state/national universities. She is associated with Ph.D. research centers. She works as an editor/reviewer for various national/international repute journals and conferences. She is the coordinator of IQAC, IPR cell, IIP cell, and research cell at the institute level. She works as Subject Chairman for various Computer Engineering subjects under Savitribai Phule Pune University (SPPU). She contributed to the SPPU syllabus content designing and revision. She received the Regional young IT Professional award from CSI in 2006. She authored the book *"Fundamentals of Data Science"* published by CRC Press (Taylor & Francis Group, US) and edited the book *"Handbook of Research on Applied Intelligence for Health and Clinical Informatics"* published by IGI Global, USA. She is a member of ISTE, ACM, CSI, IAENG, Internet Society, etc.

Anuradha D. Thakare received her Ph.D. in Computer Science and Engineering from SGB Amravati University, M.E. degree in Computer Engineering from Savitribai Phule Pune University, and BE degree in Computer Science and Engineering from Sant Gadge Baba Amravati University, Amravati, India. She works as a Professor in the Computer Engineering Department of Pimpri Chinchwad College of Engineering, Pune. Dr. Anuradha is a Secretary of the Institution of Engineering & Technology Pune LN, a Member of IEEE and ACM. She is a Ph.D. guide in Computer Engineering at SPPU, Pune. She was a General Chair of IEEE International Conference ICCUBEA 2018 and an Advisory member for International Conferences. She worked as a reviewer for the *Journal of International Blood Research*, *IEEE Transactions*, and *Scopus Journals*. She is a reviewer and examiner for Ph.D. defence for state/national universities.

She has published 78 research papers in reputed conferences and journals with indexing in Scopus, IEEE, Web of Science, Elsevier, Springer, etc. She received Research project grants and workshop grants from AICTE-AQIS, QIP-SPPU, BCUD-SPPU Pune, and Maharashtra State Commission for Women. She received the Best Women Researcher Award and Best Faculty Award from International Forum on Science, Health & Engineering. She received the best paper award in International Conferences. She delivered 20 expert talks on machine learning, evolutionary algorithms, outcome-based education, etc. She worked with industries like DRDO, NCL, etc. for research projects.

She works as Subject Chairman for various Computer Engineering subjects under Savitribai Phule Pune University (SPPU). She contributed to the SPPU syllabus Content designing and revision. She authored the book *"Fundamentals of Data Science"* published by CRC Press (Taylor & Francis Group, US) and edited the book *"Handbook of Research on Applied Intelligence for Health and Clinical Informatics"* published by IGI Global, USA.

Abbreviations

ACK	Acknowledgment
APO	Application Object
ASCENT	Adaptive Self-Configuring sEnsor Networks
BSS	Basic Service Set
CA	Cellular Automata
CCA	Cyclic Cellular Automata
CDS	Connected Dominating Set
CS	Range Carrier Sensing Range
CSMA	Carrier Sense Multiple Access
CTS	Clear To Send
DARPA	Defense Advanced Research Projects Agency
DG	Delaunay Graph
DiffServ	Differentiated Services
DIFS	Distributed Inter-Frame Space
DSN	Distributed Sensor Networks
DSSS	Direct Sequence Spread Spectrum
EOFS	Environment Observation and Forecasting System
ESS	Extended Service Set
FFD	Full Function Devices
GAF	Geographic Adaptive Fidelity
GG	Gabriel Graph
GOAFR	Greedy Other Adaptive Face Routing
GRG	Geometric Random Graphs
HEED	Hybrid Energy-Efficient Distributed Clustering
IBSS	Independent Basic Service Set
IEEE	Institute of Electrical and Electronic Engineers
INTSERV	Integrated Services
LEACH	Low Energy Adaptive Clustering Hierarchy
LLC	Link Layer Control
LMST	Local Minimum Spanning Tree
LPL	Low Power Listening
MAC	Medium Access Control
MANET	Mobile Ad Hoc Network
MDS	Minimum Dominating Set
MEMS	Micro-Electro-Mechanical Systems
MST	Minimum Spanning Tree
OSI	Open Systems Interconnection
PDA	Personal Digital Assistant
PDA	Personal Digital Assistant
QoS	Quality of Service
RCE	Random Correlated Event
RF	Radio Frequency

RN	Random Network
RNG	Relative Neighborhood Graph
RTS	Request To Send
SBYaoG	Smart Boundary Yao Gabriel Graph
SCH	Scheduling
SIFS	Short Inter-Frame Space
SIG	Special Interest Group
SSIM	Smart Sensors and Integrated Microsystems
TC	Topology Control
TDMA	Time Division Multiple Access
Tx	Range Transmission Range
UDG	Uniform Disk Graph
UWSN	Underground Wireless Sensor Network
Wi-Fi	Wireless Fidelity
WLAN	Wireless Local Area Network
WMSN	Wireless Multimedia Sensor Network
WSN	Wireless Sensor Network
YG	Yao Graph

1 Introduction and Background Study

1.1 IoT AND WSN

Wireless Sensor Networks (WSNs) have begun to fascinate researchers due to the rapid development of wireless technology and integrated devices. WSNs are often made up of small devices called nodes. These nodes contain an implanted CPU, limited computing power, and other intelligent sensors. Nodes use these sensors to monitor natural ambient elements they are humidity, pressure, temperature, and vibration. A sensor interface, processor, transceiver, and power units are typical components of a WSN node. These devices accomplish decisive tasks by allowing nodes to interconnect with one another and relay data collected by sensors. A centralized system necessitates communication between nodes. The need for this system inspires the creation of the notion of the Internet of things (IoT). The IoT idea enables quick access to environmental data. As a result, efficiency and productivity in a range of procedures improve significantly.

1.1.1 Overview of WSN

A WSN is typically defined as a network of nodes that collaborate to perceive and manage the environment around them. These nodes are joined together through wireless technology. This connection is used by nodes to communicate with one another. A typical WSN architecture comprises three components: sensor nodes, gateways, and observers (user). The sensor field is made up of sensor nodes and gateways. Interconnections between gateways and observers are made possible through dedicated networks or, more typically, the Internet as shown in Figure 1.1.

A WSN is conceptually based on a modest equation that states that Sensing + CPU + Radio = Lots of Potential [1]. A sensing unit is required to monitor the surroundings and their circumstances for instance humidity, pressure, and vibration. After the monitoring and sensing operations are completed, the CPU performs the necessary computations. Finally, computed conservational data are transmitted by Radio Unit over wireless communication channels between nodes. Finally, the data is transferred to the Gateway.

The Sound Surveillance System (SOSUS) was the first wireless network that could be characterized as a modem WSN. The US Military developed SOSUS in the 1950s to detect Soviet submarines. The SOSUS network is intended to include submerged sensors and hydrophones spread over the Atlantic and Pacific Oceans [2]. In the 1980s, the United States Defense Advanced Research Projects Agency (DARPA) pioneered the Distributed Sensor Network (DSN) effort to investigate the particular issues of adopting WSNs. The potential of DSN and its advancement in

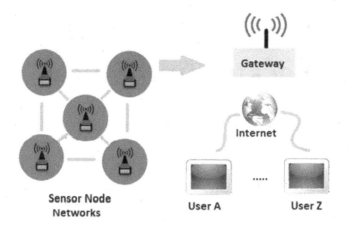

FIGURE 1.1 Wireless sensor networks (WSN).

academia have piqued the interest of academics. These issues have prompted academic and scientific researchers to investigate the potential of WSN.

For instance, IEEE has observed the subsequent fact: the low cost and high proficiencies of these miniature devices. The IEEE association has established a standard to address low data rate wireless personal area networks – IEEE 802.15.4. Based on this standard, the ZigBee Alliance published the ZigBee standard for usage in WSNs.

1.1.2 How Does WSN Works?

WSNs are made up of nodes, which are miniature computers in and of themselves. These little devices collaborate to construct centralized network systems. These networks' nodes must meet certain characteristics like efficiency, multi-functionality, and wireless capability.

Furthermore, every node in each network serves a distinct purpose. If the purpose is to accumulate data about microclimates in different sectors of a forest, for example, these nodes are located in diverse trees to form a network. This type of network should have a centralized and synchronized structure for communicating and sharing data. The sensor nodes are linked in a network with a particular topology, such as linear, star, or mesh. Any network node in any architecture has a limited broadcast range of 30 meters.

Data collection and data transfer in WSNs are achieved in four steps: data gathering, data processing, data packaging, and data transfer.

1.1.2.1 Technology

Sensor Node

It is one of the most crucial aspects of any WSN [3]. A sensor node is a small, low-power device. Despite its limited energy resources, it has a concurrent processing capability and a low cost. Figure 1.2 depicts the elements of a sensor node. Definite units of a sensor node collect and transfer data.

Introduction and Background Study

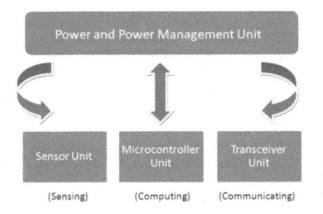

FIGURE 1.2 WSN components.

Power Source

The sensor node's base is equipped with a power source. It provides power to sensor node units such as sensing units (sensors), CPUs, and radios. Energy is required to continue performing sensing, computing, and communication operations. As a result, small sensor nodes are powered by ambient energy harvesting (AEH) techniques. Watch batteries, solar cells, and smart systems can all be used as power sources.

The AEH is accomplished in a variety of ways, including standard optical cell power generation as well as small piezoelectric crystals, micro-oscillators, and thermoelectric power generation elements [3]. All sensor nodes have limited energy supplies, and energy is necessary to fulfill all jobs. As a result, nodes may spend up to 99% of their pause time conserving energy. They only awaken to collect, transfer, and receive data.

Microcontroller

The CPU of a sensor is typically accomplished with a microprocessor and flash memory. Most of the sensor nodes have connectors that allow you to easily add exterior processing units and sensors to the main unit. Making decisions and analyzing data are two examples of crucial CPU tasks. The CPU keeps the data in flash memory until there is enough of it. Once the system has gathered sufficient data, the CPU's microprocessor unit arranges the data in envelopes since envelopes are extremely effective in data transport. These envelopes are then broadcast over the airwaves. Meanwhile, to maintain the most effective network structure, the system communicates with other nodes in the same way that it communicates with data. The CPU is associated with the base, from which it connects with the sensors and radio [4].

Sensor Transducer

A WSNs most significant component is its sensors. Sensors transform environmental variables such as light, smoke, heat, and sound, among others, into electrical impulses. Over the previous two decades, there has been rapid progress in a wide range of sensing technologies, which has facilitated sensor production:

- Gyroscopes, acoustic sensors, accelerometers, smoke sensors, magnetometers, chemical sensors, pressure sensors, and piezoelectric sensors are examples of Micro Electro Mechanical Systems (MEMS).
- CMOS sensors, including chemical composition sensors, humidity sensors, temperature sensors, and capacitive proximity sensors.
- LED sensors include chemical composition sensors, proximity sensors, and ambient light sensors.

These developments have resulted in sensors being extensively used in everyday life, particularly in sensor nodes. A distinctive node is made up of three different types of sensors: temperature, vibration, and moisture. However, certain nodes may have additional features like photographing their surroundings, detecting motion, detecting pressure, smoke detection, detecting light, and so on.

Transceiver

It is the responsibility of a sensor node's wireless communications. The four operational states of a transceiver are receive, transmit, idle, and sleep. Radio frequency (RF), infrared, and laser can all be used as wireless media in a transceiver. RF is the most popular wireless communication technology for WSNs. The typical RF operation range is 10 meters for indoors and 100 meters for outdoors.

Operating System

Operating systems used for WNSs include Tiny OS, Contiki, MANTIS, and BTunt. Tiny OS is the most open source and energy-efficient of these systems. Tiny OS employs event-driven programming instead of multithreading.

1.1.2.2 Gateways

System managers can use gateways to connect nodes to personal digital assistants (PDAs) and personal computers (PCs). There are three types of gateways: active, passive, and hybrid. The active gateway enables nodes to communicate with the gateway server in real-time. The passive gateway does not have the same freedom as the active gateway. It demands that sensor nodes give the data they need. A hybrid gateway is a grouping of these dual gates that can function in either mode.

1.1.2.3 Task Managers

They communicate with the gateways via a predefined medium, like satellite links or the Internet. Client data browsing/processing and data service are the two aspects of Task Managers. Task Managers can be thought of as a platform for information processing and retrieval. This section stores and analyses all sensor data collected. Users and administrators can access and analyses this data locally and/or remotely using an interface [5].

1.1.2.4 Communication Architecture for WSNs

A radio comprises two components: a radio transmitter and a radio receiver. These components must be present for any of the nodes for them to communicate fully with additional nodes. The radio receives information from the centralized and

Introduction and Background Study

delivers it to other sensor nodes during data transfer. The radio receives data from another node's radio and transfers it to the central system along with data reception.

The sensor node sends all data it collects to the parent node. This parent node is linked to a multi-functional computer, which permits access to other nodes' data through the user's computer interface. If the user offers instructions, they will be sent to a multi-functional computer over the Internet. These commands will be delivered to the parent node, which will then transmit the same message to its child nodes.

WSN Communication Standards and Specifications

The following are the utmost extensively used WSN communication standards:

ZigBee

The communication range of ZigBee is normally 10 meters. It can, however, transport data across large distances. This is accomplished by sending data over short distances between intermediate devices. It consumes very little power. The data rate is limited to 20 kbps.

Bluetooth

It is a wireless technology standard that enables mobile devices to communicate across short distances. The communication range is one meter to one hundred meters. It consumes very little power. The maximum data rate is 3 Mbps.

6LoWPAN

This is a method for sending and receiving IPv6 packets via IEEE 802.15.4 networks. It has a range of communication from 45 meters to 90 meters. It uses a moderate amount of power.

WSN Design Factors and Requirements

When designing WSNs, several factors must be considered. When building algorithms or protocols for WSNs, these factors have served as a guideline. Furthermore, these elements can be utilized as a criterion for comparing various systems. As a first step, application needs must be addressed before designing WSN for that application. When those criteria are identified, appropriate technologies to encounter those necessities can be chosen.

Quality of Service

Quality of service (QoS) is associated with the dependability and priority mechanisms of WSNs. As a result, sensors can accomplish several emerging applications, like object tracking and fire detection. QoS should be employed in such applications to improve WSN security and dependability. As a result, three major restrictions should be considered while developing new protocols for important applications. Data redundancy, collision, and resource limits are examples of these.

Fault Tolerance

Sensors may fail due to the hostile environment; nonetheless, this failure should not affect the WSN. Every algorithm or protocol created for WSNs would be fault-tolerant in this environment. Each application has a varied amount of fault tolerance. In household applications, such as humidity or temperature monitoring, high fault tolerance is not required because sensors are not easily destroyed. Outdoor habitats, on the other hand, are classified as harsh environments. To avoid the possibility of failure, high fault tolerance is essential.

Time of Data Delivery

Delays in data delivery are limited in applications that demand real-time delivery. In time-critical applications, service latency should be limited. In healthcare, for example, if doctors do not receive signals on time, patients' lives may be risked. As a result, when developing protocols or algorithms, the time elapsed between the source and attention must be considered. It is critical to adhere to a minimum allowable delay, which is determined mostly by the type of the application. Scalability: Because hundreds or thousands of sensors are positioned based on the application, the designer must exercise caution when dealing with the prospect of network expansion. This high density of sensors, however, must be used to cover as wide an area as feasible.

Energy Consumption

In some situations, replacing or recharging batteries may be impossible or complicated. As a result, battery life has a significant impact on the lifetime node. As a result, the overall network's lifespan will be shortened. In the worst-case scenario, when nodes are routers, this failure will damage the entire network. The major tasks that sensors need energy for include sensing, processing, transmitting, and receiving. Furthermore, noise might increase power usage owing to retransmissions. WSN-specific data compression algorithms are being studied to reduce power consumption. Here study determined that data communication munches more resources over data processing. Several communication activities take place in WSNs, with the transmission, reception, frequency synthesizers, voltage management, and so on, and all of these functions drain power from sensors.

Gathering Data

Based on how data is acquired, WSN applications are categorized as Event Detection (ED) or Spatial Process Estimation (SPE). In the emergency department, a specific occurrence, such as a fire, must be identified by positioning sensors. SPE, on the other hand, is used to forecast a physical situation, like ground temperature in a volcano or air pressure. Though, some environmental applications may fall within both categories.

Communication Architecture

Sensors have two basic purposes. These tasks detect data or route it to the sink node. The sink node, like all other nodes in the network, communicates in a tiered fashion. A protocol stack is a type of communication architecture.

Homogeneous vs. Heterogeneous

When all sensors in a WSN are the same, it is referred to as a homogenous network. On the other hand, heterogeneous networks are made up of several kinds of sensors.

Though homogeneous networks are simple to maintain, heterogeneous networks might offer a better resolution due to their diverse energy models. Because certain nodes have more energy than others, it is sometimes necessary to allocate a substantial duty to them. Cluster heads serve as routers in a network of nodes. As a result, it is advantageous to have more energy to transmit data than the other nodes. As a result, heterogeneous networks may be able to maximize network longevity. Nonetheless, homogenous networks are simple to set up. Furthermore, cluster heads can be switched to avoid node death.

1.1.3 Security Issues in WSN

WSN security is vital, especially if they perform mission-critical activities. For example, in a health care application, the privacy of a patient's health record should not be disclosed to third parties. Attacks on the security of WSNs are divided into two types: active and passive. In active attacks, the attacker harms the functioning of the targeted network. This could be the attacker's main goal, which can be simply noticed as related to passive attacks. Active attacks are classified as hole attacks (black holes, sinkholes, wormholes, and so on), denial-of-service (DoS) attacks, jamming attacks, flooding attacks, and Sybil attacks. Passive attackers are typically physically disguised and either tap the data link to acquire data or destroy any network operational equipment. Eavesdropping attacks, node tampering attacks, node malfunctioning attacks, node demolition attacks, and traffic analysis attacks are all examples of passive attacks.

To assess safety for WSNs, steps may be accomplished toward attacks, similar to in another network: intrusion prevention and intrusion detection. Intrusion detection and prevention strategies are the first lines of protection against invaders. However, like with any protection machine, intrusions cannot be avoided. The assault and compromise of a single node can bring about the disclosure of essential community protection statistics to intruders. As a result, the preventive protection system fails. As a result, Intrusion Detection Systems (IDSs) are intended to detect intrusions before they reveal information about the protected system resources. IDSs are continually seen as a subsequent line of defense in terms of safety. IDSs are the Internet correspondent of burglar alarms, which are employed in today's physical safety systems.

Wireless network systems are predicted to be extensively used as a result of the development of Micro-Electrical Systems (MEMS). MEMS is an extremely small-scale combination of electrical devices and mechanical structures. Various surveys would be directed before MEMS may be used in WSNs. The impacts of extremely high node density would be examined. The increased use of WSN devices, also the anticipated difficulty in retrieving specific devices throughout the whole network, would not be overlooked.

Furthermore, IoT is expected to have a massive effect on our lives withinside the close to future. WSNs can be included in IoT, and a plethora of sensor nodes will hook up with the Internet. They will work with different nodes to sense and evaluate their surroundings.

1.2 IoT AND SENSOR NETWORK APPLICATIONS

Sensors are classified into five types. They are intended for use in underground, underwater, terrestrial, multimedia, and mobile applications. Although terrestrial sensors are less expensive, their battery capacity is limited. Burying sensors are sometimes essential to evaluate particular circumstances in applications like agriculture or mining. As an outcome of these applications, the cost of sensors may be high. Acoustic waves, on the other hand, are used for communication in underwater applications. Sensors are unreachable due to the hostile environment of the water. As a result, the sensors' energy usage should be carefully evaluated [6].

To track objects or actions in digital applications, mics and cameras are required. Furthermore, to forward and receive video, audio, or pictures need an efficient data rate. Finally, portable sensors are used in mobile applications such as in the military field. In this case, pay close attention to the communication range. Sensors in the network can move and change places. To transfer data and structure the network, dynamic methods are required. The following are some IoT with WSN applications:

1.2.1 WIDE SPACE APPLICATIONS

1.2.1.1 Smart Cities

It is becoming more popular these days. The key components of such a city are intelligent items with a CPU and a transceiver device to connect with another [7]. These intelligent items have the potential to create a secure and intelligent environment. This situation is also known as the IoT. The Internet of vehicles (IoV) is a subclass of IoT that will make transportation systems more sophisticated. There are three possible ways for communication in IoV like vehicles to vehicles (V2V), vehicles to infrastructure (V2I), and infrastructure to infrastructure (I2I) [8]. VANETs are utilized in a variety of applications, including identification of vehicle speed, avoidance of traffic jam, optimal route, and V2V communication. IoVs are made up of several volumes of vehicles and some wayside stations that can be used for remote operations in an ad hoc or cellular fashion. Vehicles' movement, orientation, and unstable topology are heavily constrained.

1.2.1.2 Smart Environmental

Environmental monitoring has a long history. Another broad area of application is pollution monitoring. Many articles have advised using WSNs in this event due to the need of having spotless environments with as diminutive pollution as feasible [9]. When outdoor sensors are installed in a big region, fire detection systems have been examined. The most common environmental applications like detections of a flood, volcano, earthquake, and chemical hazardous. Furthermore, underwater applications like water pollution prevention and sea mammal monitoring are possible. Furthermore, sensors close to the animals' bodies have been fitted to measure conditions linked to rear conditions, generated gases, and animal temperature.

Introduction and Background Study

1.2.1.3 Smart Agricultural

Wireless sensors are having a big impact on agricultural applications. Monitoring soil and crops have gotten a lot of interest from experts working on everything from farmed irrigation to fertilizer organization. Traditional methods of assessing agricultural metrics may be problematic, exclusively in big arenas. As a result, sensors are an appropriate choice for data collection. WSN is superior to the traditional methods that require effort and attendance [10]. As a result, low-value intelligent sensors with small batteries and wi-fi conversation abilities were demonstrated. In farmed applications, two types of sensors are used terrestrial sensors and subsurface sensors [11].

1.2.1.4 Defense Applications

An important feature of Defense applications is the ability to broaden the scope of interest from information collection to monitoring or surveillance. The defense applications are divided into four categories they are battlefield, force protection, urban warfare, and non-violent conflict.

1.2.2 SMALL SPACE APPLICATION

1.2.2.1 Operational Conditions Monitoring

Buildings and bridges must be mechanically inspected following natural calamities such as earthquakes [12].

1.2.2.2 Industrial Applications

The major crucial characteristic that distinguishes the automata industries is the control of industrial machines. Furthermore, manufacturing monitoring is a crucial procedure in industrial applications [12].

1.2.2.3 Healthcare Applications

Sensors on the body that measure health conditions such as respiration, blood pressure, blood flow, ECG, and oxygen. Based on the data obtained by the intelligent sensors, new medications could be designed.

1.2.2.4 Intra-Vehicle Applications

Smart automobiles are equipped with a plethora of sensors. These sensors provide a clear picture of a vehicle's status. In some applications, the most important characteristics are pressure, engine status, and speed. Container monitoring aboard trains or ships is another application for intra-vehicle sensors. These applications do not require energy awareness. However, energy usage is still taken into account to limit CO_2 emissions in the environment. Indeed, most communication systems may contribute just a minor amount to CO_2 emissions [13].

1.3 OSI AND IoT LAYER STACK

IoT protocol stack layers outline the operations of IoT stack Layers 1, 2, 3, 4, 5, and 6, as well as IoT Layer 7. IoT contains protocol levels ranging from 1 to 7, similar to

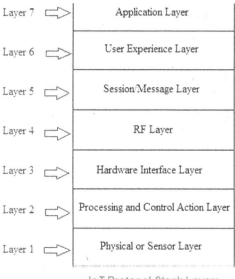

FIGURE 1.3 IoT Protocol Stack Layers.

other wired and wireless technologies. We've seen the OSI (open system interconnection) model, which outlines a seven-layer networking architecture. OSI outlines the functions and interfaces of each of these layers. Let's have a look at the functional description of the seven tiers of the IoT protocol stack, as depicted in Figure 1.3.

1.3.1 Physical or Sensor Layer

This IoT Layer 1 interfaces with physical components in the same way that the OSI physical layer does. Sensors such as humidity sensors, temperature sensors, pressure sensors, heartrate sensors, pH sensors, smell sensors, and so on are among the physical components. The sensors are used to detect various factors depending on the application. Because there are numerous sensors available for the same functionality, the suitable sensor is chosen depending on cost and quality. This Layer-1 is responsible for delivering sensed data to the IoT stack for further processing.

1.3.2 Processing and Control Layer

This layer processes the data provided by Layer-1 via sensors. At this tier, the microcontroller/processor and operating system are critical. Various development kits, including Arduino, NodeMCU (based on ESP32 or ESP8266), ARM, PIC, and others, can be used for this purpose. Android, Linux, IOS, and other popular operating systems are used.

Introduction and Background Study

1.3.3 Hardware Interface Layer

This layer contains communication components or interfaces such as RS232, RS485, SPI, I2C, CAN, SCI, and so on. These interfaces are used for serial or parallel communication in synchronous/asynchronous modes at various baud rates. The interface protocols specified above enable seamless communication.

1.3.4 RF Layer

This radio frequency layer contains RF technologies based on short or long-range, as well as the data rate required by the application. Wifi, Bluetooth, Zigbee, Zwave, NFC, and RFID are examples of common indoor RF/wireless technologies. GSM/GPRS, CDMA, LTE-M, NB-IoT, and 5G are some of the most common outdoor RF cellular technologies. The RF layer communicates data via radio frequency-based EM waves. Another type of data communication method employs light waves. This type of light-based data transfer is known as LiFi.

1.3.5 Session/Message Layer

This layer handles message protocols such as MQTT, CoAP, HTTP, FTP (or Secured FTP), SSH, and others. It specifies how communications are sent to the cloud. Refer to the architectures of the MQTT protocol and the CoAP protocol.

1.3.6 User Experience Layer

This layer is concerned with giving the greatest experience possible to end-users of IoT products. To do this, this layer is in charge of creating sophisticated UI designs with a plethora of functionality. For the design of GUI interface software, various languages and tools have been developed. These technologies include object-oriented and procedure-oriented technologies, as well as database languages (DBMS, SQL) and analytics tools.

1.3.7 Application Layer

This layer makes use of the remaining six levels to create the required application. The following are examples of typical IoT case studies or applications: Smart Homes, Smart Cities, Smart Agriculture Farming, and so on.

1.4 PROTOCOLS IN WSN AND IoT

Routing protocols for WSNs, IoT, and AdHoc wireless networks are covered in this section. There are various classifications or types of routing protocols that are based on protocol operation or functionality as well as a network structure. Routing protocols are classified as reactive or proactive. Reactive routing is also known as on-demand routing since the protocol only looks for a route to a destination when it is required. Periodic messages are used in proactive routing to deliver messages to nodes about their surroundings, ensuring that they always have a route to their destinations.

1.4.1 Routing Protocol for Low-Power and Lossy Networks

Routing protocol for low-power and lossy networks (RPL) is a source routing protocol that is based on IPv6 and is independent of the link used for routing. It was designed for low-power, lossy networks, and was standardized by the IETF in 2011. It is commonly regarded as the de-factor routing protocol for the IoT.

A protocol is said to be a distance-vector if its nodes can manipulate distance vectors or arrays to other nodes in the network. This signifies that the protocol's nodes interact with one another within the domain. To have an effective interaction between nodes, computation complexity, and message overhead must be kept to a minimum, and each node must notify other nodes of any changes in topology. The network topology is the pattern of organization in which nodes in a network are connected. A distance-vector protocol always computes the direction (the next hop's address) and distance (the cost to reach a node) to any node in the network. Every node maintains a vector with the shortest distance to every other node.

1.4.2 Cognitive RPL

1.4.2.1 Cognitive and Opportunistic RPL

Cognitive RPL (CORPL) is an extension of RPL developed for cognitive networks – a network with a perceptive process that can notice current network conditions, act on them, and self-learn as a result of its actions. CORPL, like RPL, employs DODAGs, but with some variations.

CORPL was designed to exploit the DAG in the same way as RPL does but with an opportunistic approach. There are two primary processes in CORPL: selecting a forwarder set and selecting a unique forwarder. The first phase is for each node in the network to choose as many next-hop neighbors as feasible. In the second stage, the nodes use a coordination system to find the best receiver among the selected forwarder set; after the best receiver is determined, the node will enable it to forward the necessary packets. Each node maintains a set of forwarders from which the next-hop/forwarding node is chosen opportunistically. According to, using opportunistic forwarding in CORPL enhances end-to-end throughput and reliability by utilizing the inherent features of the wireless channel – which is a major concern in lossy networks.

1.4.3 Lightweight On-Demand Ad hoc Distance Vector Routing – Next Generation (LOADng)

LOADng is a reactive protocol that was created to provide efficiency, scalability, and security in LLN routing; it is a lightweight distance-vector protocol. It does not maintain a routing table for different nodes, but instead works on-demand, initiating a route discovery whenever there is a need to transfer packets to a destination node;

Introduction and Background Study

as explained in previous sections, reactive protocols have lower routing overhead and memory consumption than proactive protocols.

LOADng, like other reactive routing protocols, uses three separate messages: route request (RREQ), route reply (RREP), and route error (RERR). When there is a requirement for packet delivery to discover a path to the target node, the sender node sends RREQ; the destination node responds with an RREP after receiving the RREQ from the sender. When a link fails or is broken, the destination sends RRER to the original sender of the packets it is receiving. Route discovery, path maintenance, and path metrics are the three protocol processes involved in LOADng.

1.4.4 Collection Tree Protocol

This routing protocol was designed particularly for WSNs. It is a routing technique based on distance and vectors. Collection tree protocol (CTP) was the TinyOS defactor routing standard before the development of the RPL. It is widely regarded as a universal reference protocol for WSNs.

CTP builds and maintains a tree-based topology utilizing routing messages, often known as beacons, which report data messages to the sink, the network's root. CTP employs the adaptive beaconing approach to ensure that routing messages are sent to the root.

1.4.5 Channel-Aware Routing Protocol

Channel-aware routing protocol (CARP) is a routing protocol for underwater WSNs. It makes use of multi-hop data transmission to the underwater WSN's sink. It is a cross-layer protocol that uses connection quality information to calculate cross-layer delay. CARP picks nodes with a recent history of successful transmissions to their neighbors based on information about link quality. To connect voids and shadow zones, the protocol combines link quality with hop count, which is simple topology information. It is also capable of selecting robust links by utilizing power control.

During the network configuration, the root node broadcasts a HELLO message to all nodes in the network. Every node can obtain its hop count – distance to the node – using the broadcast message, which is extremely useful. Whenever a packet needs to be transferred, the sender node chooses the best relay to the destination node using PING and PONG messages. The PING message is sent by the node to initiate a packet transfer, and the PONG message is sent by any node that receives the PING message and forms a relay to the destination node.

Time is recorded during the exchange of PING and PONG messages to obtain a relay. Along with the time, goodness is computed for each node, and the good value is then used to calculate the link quality of all possible relays to the destination node. To transfer the packets, the relay with the best link quality is chosen. While sending the PING messages, the power used to send them is also computed, allowing CARP to use power control to select robust links for packet transfer.

1.4.6 E-CARP

This protocol extends CARP to support greedy and location-free hop-to-hop routing to ensure energy-efficient packet forwarding from sensor nodes to sink. CARP ensures that data acquired by sensor nodes is not ignored by their presence in the network; however, unwanted forwarding may occur from those nodes in the network; E-CARP is designed to address this issue by enabling caching of sensory node data at the sink.

Another feature that is underutilized in CARP is the reusability of network relays. There is usually no need for a PING-PONG message transfer between nodes when the network is stable. E-CARP is designed to capitalize on the reusability of previous links by prioritizing previously used links before initiating a transfer.

1.5 ENERGY CONSUMPTION AND NETWORK TOPOLOGY

The IoT is currently being developed, in which devices or things are linked over the IP protocol stack. Most of these devices have restricted hardware resources, such as low processing power, limited memory, low power, and inadequate communication abilities like short-range, low bitrate, and short frame size. The most frequent communication methods for IoT are IEEE 802.15.4, low power Wi-Fi, and these sorts of networks are known as low power and lossy networks (LLNs) [14]. The IPv6 routing protocol for LLNs was established in 2012 by the IETF ROLL working group to standardize the routing process for LLNs. RPL design is complex and differs from earlier routing ideas due to the inherent LLNs of low data rates, a high probability of node and connection failures, and limited energy resources [15].

The first order constraint of IoT is energy consumption, and evaluating the energy consumption of an IoT operating RPL is required. The IoT's energy consumption comprises both the energy consumption of an individual node and the energy consumption of the whole system. An individual node can operate in five modes: complete function, small energy, forwarding, eavesdropping, and receiving. The complete function process consumes the most energy. The transmission distance of radio frequency signals is also associated with the energy consumption of a single node. The IoT primarily relies on multi-hop data forwarding; nodes near the root node take the data furtherance of other nodes, and the energy consumption is more, resulting in unbalanced energy consumption of the system.

1.6 CHALLENGES FOR ENERGY CONSUMPTION IN ioT NETWORKS

Many IoT components will need to be basic and able to work reliably and independently for extended periods. However, more complex components, such as data aggregation points and gateways between networks of connected devices and the outside world, are also required. Aside from all of the benefits of IoT for energy savings, positioning IoT in the energy sector poses hurdles. The succeeding section discusses the difficulties and prevailing options for implementing IoT-based energy systems.

Introduction and Background Study

1.6.1 Energy Consumption

The primary focus of IoT platforms in energy systems is energy conservation. A huge volume of IoT devices broadcast data in energy systems to enable communication via IoT. A significant quantity of energy is required to run the IoT system and transport the massive volume of data created by IoT devices [16]. Various ways, however, have been taken to lower the power consumption of IoT systems. For instance, you could configure the sensors to go into idle mode and work as and when required. It has been broadly evaluating about how to design efficient communication protocols that allow distributed computing approaches that enable energy-efficient communications. Radio optimization strategies including modulation optimization and cooperative communication were proposed as a probable solution. Furthermore, energy-efficient routing solutions like cluster designs and the use of multi-path routing algorithms were recognized as another alternative [17].

1.6.2 Combination of IoT with Subsystems

Integration of an IoT system with energy subsystems is a significant issue. Because each subsystem in the energy sector is unique, it employs a wide range of sensor and data communication technologies. As a result, methods for co-ordinating data transfer among IoT-enabled energy system subsystems are required [18]. Modeling an integrated framework for the energy system is a strategy for addressing the integration challenge by taking into account a subsystem's IoT requirements. Other proposals argue for the development of co-simulation models for energy systems to integrate the system and reduce synchronization delay error between subsystems [19].

1.6.3 User Privacy

Individual or cooperative energy users have the right to preserve the privacy of their data when it is exchanged with an association, which is known as privacy. As a result, appropriate data access like the number of energy users and the quantity and kinds of energy-consuming equipment is incredible. Indeed, the type of data collected by IoT allows for better decision-making, which can inspire energy production, distribution, and consumption [20]. To reduce the invasion of users' privacy, energy providers should acquire user agreement before using their information, ensuring that the users' data is not shared with third parties. A reliable privacy management system in which energy users have control over their data and secrecy is also recommended [21].

1.6.4 Safety Challenge

From production, transmission, and distribution to consumption, the usage of IoT and the integration of communication technologies in energy systems increase the potential of cyber-attacks on users' and energy systems' information [22]. These threads illustrate the uncertainty confronting the energy sector. Furthermore,

IoT-based energy systems are broadly positioned in large geographical areas to deliver energy-related services. They are increasingly exposed to hackers as a result of the broad implementation of IoT strategies. A study offers an encryption strategy to protect the information of energy from cyberattacks to address the issue. Furthermore, distributed control systems that enable control at different IoT system levels are recommended to lessen the possibility of cyberattacks and boost system security [23].

1.6.5 IoT Standards

IoT connects a single device to an enormous amount of devices through the use of various technologies and protocols. Unpredictability amongst IoT devices that implement several criterions introduces a new contest. In IoT facilitated systems, standards are divided into two categories like network and communication protocols, and other is data-aggregation standards, as well as regulatory requirements relating to data safety and secrecy. Among the problems challenging the assumption of criteria in IoT to comprise criterions for processing amorphous data, safety and privacy concerns, as well as regulatory requirements for data marketplaces [24]. One approach to acquainting the problem of standardization of IoT-based energy systems is to define a system with shared knowledge that all users may access and use equally. Another possibility is for parties to collaborate to develop open information models and standards-based protocols. As a result, the public will have access to standards that are free and open [25].

1.6.6 Architecture Design

IoT facilitated systems are comprised of many technologies, as well as an accumulative number of smart networked devices and sensors. The IoT is expected to enable communications at any time and from any location for any associated services, in general, in a self-directed and ad hoc manner. This infers that the IoT systems are intended with complicated, decentralized, and mobile features dependent on their application requirements [25]. Given the differences in characteristics and requirements, a reference design cannot be a single resolution for all IoT applications. As a result, IoT systems require diverse reference structural designs that are uncluttered and adhere to criterions. Furthermore, the structural design should not bounds users to utilize secure and end-to-end IoT communications [26].

1.7 SUMMARY

This section describes the introduction and background details of IoT and WSN and their applications in different areas. Also discussed is the IoT layer protocol stack. Furthermore, addressed the various protocols being used in WSN and IoT with energy consumption and network topology. At last, discussed various challenges for energy consumption in IoT networks.

Exercise
1. What is the significance of IoT and WSN?
2. Explain in detail the applications of IoT and WSN.
3. Explain the IoT layer stack in detail.
4. List and discuss in detail the protocols used in WSN and IoT.
5. What are the different challenges for energy consumption and IoT networks?

REFERENCES

[1] Ephrem, E. Architecture of Wireless Sensor Networks. Retrieved October 8, 2015, from http://servforu.blogspot.com.tr/2012/12/architecture-of-wirelesssensor-networks.html

[2] Wang, Q. and Balasingham, I. 2010. Wireless sensor networks – An introduction. *Wireless Sensor Networks: Application-Centric Design*, 1–15. DOI: 10.5772/13225

[3] Yinbiao, D. and Lee, D. 2014. IEC White Paper Internet of Things: Wireless Sensor Networks. *International Electrotechnical Commission White Paper*.

[4] Raja, C. and Sinnaiya, M. 2011. Analysis and report of wireless sensor networks. *International Journal of Computer Science*, 8:1–2.

[5] Villegas, M. A. E., Tang, S. Y. and Qian, Y. 2007. Wireless Sensor Network Communication, Architecture for Wide-Area Large Scale Soil Moisture Estimation and Wetlands Monitoring, Department of Electrical and Computer Engineering, University of Puerto Rico at Mayaguez.

[6] Mahdi, M. A. and Hasson, S. T. 2017. A contribution to the role of the wireless sensors in the IoT Era. *Journal of Telecommunication, Electronics and Computer Engineering*, 10:1–6.

[7] Botta, A., Donato, W., Persico, V. and Pescapé, A. 2016. Integration of cloud computing and Internet of Things: A survey. *Future Generation Computer Systems*, 56:684–700.

[8] Keertikumar, M., Shubham, M. and Banakar, R. M. 2015. Evolution of IoT in smart vehicles: An overview. in 2015 International Conference on Green Computing and Internet of Things (ICGCIoT).

[9] Xiaojun, C., Xianpeng, L. and Peng, X. 2015. IOT-based air pollution monitoring and forecasting system. Computer and Computational Sciences (ICCCS), 2015 International Conference on IEEE.

[10] Kaewmard, N. and Saiyod, S. 2014. Sensor data collection and irrigation control on vegetable crop using smart phone and wireless sensor networks for smart farm. Wireless Sensors (ICWiSE), 2014 IEEE Conference on IEEE.

[11] Zenglin, Z., Pute, W., Wenting, H. and Xiaoqing, Y. 2012. Design of wireless underground sensor network nodes for field information acquisition. *African Journal of Agricultural Research*, 7:82–88.

[12] Hodge, V. J., O'Keefe, S., Weeks, M. and Moulds, A. 2015. Wireless sensor networks for condition monitoring in the railway industry: A survey. *IEEE Transactions on Intelligent Transportation Systems*, 16-3:1088–1106.

[13] Darwish, T., Bakar, K. A. and Hashim, A. 2016. Green geographical routing in vehicular ad hoc networks: Advances and challenges. *Computers & Electrical Engineering*, 64:436–449.

[14] Winter, T., Thubert, P., Brandt, A., Hui, J., Kelsey, R., Levism, P., Pister, K., Struik, R., Vasseur, J. and Alexander, R. 2012. RPL: IPv6 Routing Protocol for Low-Power and Lossy Networks, RFC 6550(Protocol Standard), Mar.

[15] Wang, Z. M., Li, W. and Dong, H. L. 2018. Analysis of energy consumption and topology of routing protocol for low-power and lossy networks. *Journal of Physics Conference Series*, 1087(5):052004. DOI: 10.1088/1742-6596/1087/5/052004

[16] Kaur, N. and Sood, S. K. 2015. An energy-efficient architecture for the Internet of Things (IoT). *IEEE Systems Journal*, 11:796–805.

[17] Lin, Y., Chou, Z., Yu, C. and Jan, R. 2015. Optimal and maximized configurable power saving protocols for corona-based wireless sensor networks. *IEEE Transaction Mobile Computing*, 14:2544–2559.

[18] Shakerighadi, B., Anvari-Moghaddam, A., Vasquez, J. C. and Guerrero, J. M. 2018. Internet of Things for Modern Energy Systems: State-of-the-Art, Challenges, and Open Issues. *Energies*, 11:1252.

[19] Kounev, V., Tipper, D., Levesque, M., Grainger, B. M., Mcdermott, T. and Reed, G. F. 2015. A microgrid co-simulation framework. In Proceedings of the 2015 Workshop on Modeling and Simulation of Cyber-Physical Energy Systems (MSCPES), Seattle, WA, USA. 1–6.

[20] Porambage, P., Ylianttila, M., Schmitt, C., Kumar, P., Gurtov, A. and Vasilakos, A. V. 2016. The quest for privacy in the internet of things. *IEEE Cloud Computing*, 3:36–45.

[21] Jayaraman, P. P., Yang, X., Yavari, A., Georgakopoulos, D. and Yi, X. 2017. Privacy preserving Internet of Things: From privacy techniques to a blueprint architecture and efficient implementation. *Future Generation Computing Systems*, 76:540–549.

[22] Poyner, I. and Sherratt, R. S. 2018. Privacy and security of consumer IoT devices for the pervasive monitoring of vulnerable people. *In Proceedings of the Living in the Internet of Things: Cybersecurity of the IoT—2018, London, UK*, 28–29 March 2018; 1–5.

[23] Roman, R. and Lopez, J. 2012. Security in the distributed internet of things. *In Proceedings of the 2012 International Conference on Trusted Systems, London, UK*, 17–18 December 2012; 65–66.

[24] Banafa, A. IoT Standardization and Implementation Challenges. 2016. Available online: https://iot.ieee.org/newsletter/july-2016/iot-standardization-and-implementation-challenges.html (accessed on 10 May 2019).

[25] Chen, S., Xu, H., Liu, D., Hu, B. and Wang, H. 2014. A Vision of IoT: Applications, Challenges, and Opportunities with China Perspective. *IEEE Internet Things Journal*, 1:349–359.

[26] Al-Qaseemi, S. A., Almulhim, H. A., Almulhim, M. F. and Chaudhry, S. R. IoT architecture challenges and issues: Lack of standardization. *In Proceedings of the 2016 Future Technologies Conference (FTC), San Francisco, CA, USA*, 6–7 December 2016; 731–738.

2 IoT and Topology Control: Methods and Protocol

2.1 SENSOR NETWORK TOPOLOGIES

Typically, sensors, smart sensors, and sensor systems combine sensing, processing, communication, and power subsystems into a single integrated system. While sensors can be used in isolation for specific applications, multiple sensors are commonly integrated into higher-level topologies to deliver real-world applications. These topologies can vary in complexity from a single node connected to an aggregator to fully meshed networks distributed over a large geographical area. Sensor topologies can also be described as having either a flat or a hierarchical architecture. In a flat (peer-to-peer) architecture, every node in the network (sink node and sensor node) has the same computational and communication capabilities. In a hierarchical architecture, the nodes operate close to their respective cluster heads. Hence, nodes with lower energy levels simply capture the required raw data and forward it to their respective cluster heads. Usually, the cluster heads possess more processing and storage capacity than any ordinary sensor node.

2.1.1 STAR NETWORK (SINGLE POINT-TO-MULTIPOINT)

The star topology includes one base station. The base station shall transmit the data received by the network node. This sort of network topology has one base station for the data that sends and receives the network nodes. The base station is centrally positioned on the network. In the star topology, all the nodes are connected to the base station for data transportation. This is seen in Figure 2.1.

In star topology, if the network can transmit data to a node, it is then transmitted via the base station. No topology node may send or receive information directly from another node. Remote nodes can send and receive messages from a single base station. This is a very useful wireless sensor network because the sensor node will maintain a minimum electricity consumption. The sensor and base station use very little communication time in this type of network. The limitation of this network is that the base station must be within the range of the radio transmission of all nodes. The star topology network is not as robust as any other network topologies. In this type of network, when the base station fails, the whole network will fail, which is the main drawback of this type of network.

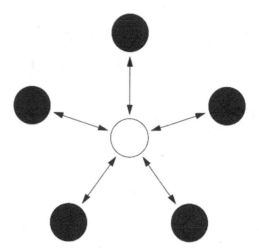

FIGURE 2.1 Star network topology.

2.1.2 Mesh Network Topology

All nodes are linked to one another in this type of network. Any node can directly send information to any node. If any nodes are disconnected from the range of the network sensor, the issue of data transportation is not faced by other nodes. This form of the network enables multihop communication, i.e. if any node wants to send data to another node, and at that time, this node is not accessible in the radio communication range, then another intermediate node is required to send data to that specific node. The strengths of this network are reliability and scalability. We should be able to increase our network size in this form of network, which has no range of nodes.

The power consumption of the sensor node indicates the drawback of this system because it supports the multihop communication system. Sensor nodes have very little battery capacity, which is why multihop communication would lose easily compared to traditional node-to-node communication. It also increases the arrival time of the message as the number of contact hops increases at the destination. An example of this type of topology is shown in Figure 2.2.

2.1.3 Hybrid Star-Mesh Network Topology

The hybrid mesh network offers a stable and multipurpose communication network. This kind of network will consume the minimum power of the sensor nodes. When a sensor node has less battery power, it does not send messages to the other nodes in this type of network. Other nodes in the networks can communicate and can send messages to sensor nodes that have low power. Multihop communication nodes have high power and, if possible, they can also be plugged into electrical switches. This topology has been implemented by the new Zigbee. An example of hybrid topology can be seen in Figure 2.3.

IoT and Topology Control

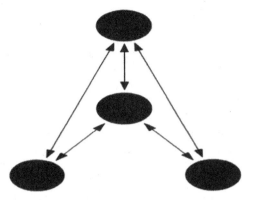

FIGURE 2.2 Mesh network topology.

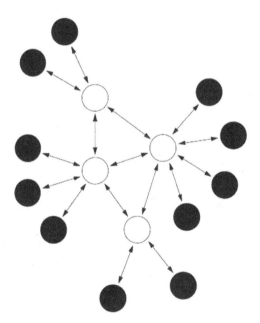

FIGURE 2.3 Hybrid star mesh network topology.

2.2 IoT AND TOPOLOGY CONTROL METHODS

Topology management is an effective strategy used by WSNs to achieve energy efficiency and longer network life without compromising important network performance, such as connectivity and throughput. The concept of topology control is to allow sensor nodes a sense of control over certain parameters so that these parameters can be controlled in a way that supports the network. In particular, the sensor nodes have the capacity to change the transmitting scope of their radio, to turn to the different modes of service, or even to settle on the eligibility of the nodes to access the network backbone. These characteristics are the parameters used to achieve energy efficiency and extend network life by implementing reduced topology.

In WSNs, a topology gives information about a group of nodes and connections between a set of nodes. Each sensor node senses its neighbors and its relative connections using maximal transmitting power to create a network topology. The node will make decisions to create a network based on the information received. The downside to this approach is that it may either be too dense (sensitive to unnecessary interference) or too sparse (sensitive to network partitioning) [1]. This approach is also very important. A proper topology control should be used to remove unwanted network connections without losing network efficiency to prevent this issue.

Topology control has two primary purposes. The first goal is to conserve resources and prolong the lifetime of the sensor node and network. The first objective is *topology controls* to provide a framework for the change in the propagation spectrum of sensor nodes, eventually reducing the energy usage during transmission. As an outcome, communication over long distances is dropped, while communication over the short distance is picked. From the point of view of energy consumption, short-distance direct communication is more energy-efficient than long-distance communication [2]. Reducing transmission power will therefore eliminate long-distance linkages that can waste energy resources. The second goal is to overcome collisions. Apart from discarding inefficient connections, the use of a small transmission range successfully eliminates long-distance nodes, resulting in a sparse network. The results of this include a decrease in packet retransmission and interruption and an increase in the performance of the network.

Topology control can be carried out in three ways. As discussed earlier, minimizing the power produced during transmission by changing the transmission range of the wireless sensor node radio is a typical method. In addition, sensor nodes that are sitting idle and do not engage in transmitting and receiving can switch off their radios or can shift to sleep mode. This strategy will have substantial energy savings as the energy consumption during idle mode is very significant relative to the energy consumed during sleep mode [3]. Topology control can ultimately be performed by a clustering technique. Based on the selection criterion, the sensor nodes will choose a group of nodes to form a cluster. This provides control over the topology to achieve energy conservation and allows for balanced hierarchical network architecture. The possible selection parameters are residual energy, the number of adjacent nodes, or the node identifier. In clustering, data forwarding and aggregation operations are allocated to the nodes in the collection to minimize the amount of packet retransmission and optimize energy resources.

According to the energy-saving methodology, classified topology control algorithms have been spread. For this criterion, the method groups the topology control algorithms into four categories, as seen in Figure 2.4. The four types are power adjustment, power mode, clustering, and hybrid. Power adjustment is a strategy that decreases the energy consumption of the WSN by manipulating the transmission power of the nodes. In contrast, power mode saves resources by shutting off idle node radios and putting the nodes in sleep mode. The third group, known as clustering approaches, conserves resources by objectively choosing a collection of neighbor nodes to create the energy-efficient backbone of the network. Finally,

IoT and Topology Control

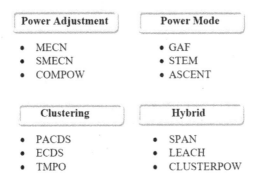

FIGURE 2.4 The four classifications of 2D-distributed topology control algorithms and 12 representative algorithms.

hybrid approaches further boost energy savings by combining a clustering approach with either power mode or power adjustment method.

2.2.1 POWDER ADJUSTMENT APPROACH

The power adjustment method helps nodes to vary their transmitting power in order to reduce the energy produced by the transmission. Instead of communicating at full transmission capacity, nodes work collaboratively to change and locate the necessary transmission power to form a connected network. The following sections describe the three algorithms for power adjustment.

1. **Minimum Energy Communication Network (MECN):** Rodoplu et al. [4] suggested a localized and position-based algorithm that minimizes the energy involved in the transmission of WSN packets. The idea of this algorithm is to create a topology consisting of the lowest energy paths to be transmitted from any wireless network sensor to a sink node using the principle of "relay transmission".

The MECN algorithm works in two steps. In the first step, each node can find its neighbor set. Typically, the neighbor set of a node includes all the nodes within its field of communication. Here, the node adds to its neighbor collection only those that can communicate directly by investing minimal packet transmission power. In other words, a node can only admit another node to its neighbor set if:

- It can connect directly with this node, and
- There is no other way to communicate with this node by using relays and wasting less energy on transmitting than direct communication.

Figure 2.5 shows this distinction. Algorithm 1 defines the building mechanism set by the neighbor.

In the second step, the nodes run the Bellman-Ford shortest path algorithm to evaluate the minimum energy path to the node of the sink. Each node communicates

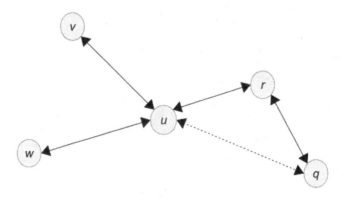

FIGURE 2.5 Neighbors of the node u: It can directly exchange packets with v, w, r, and q, but discovers that relayed packet transmissions to q via r are more efficient than direct communications. So, it does not include q in its neighbor set.

the cost of using itself as a relay to the sink (here, the cost is the minimum power consumption required to send a packet to the sink). When a node u receives cost information from a neighbor node v, it determines the minimum cost of the path to the sink relayed by v as in 2.1,

$$\text{Cost}(u, v) = \text{Cost}(v) + d(u, v)^n + \beta \qquad (2.1)$$

where $d(u, v)$ is the Euclidean distance between the nodes u and v (it is assumed that the nodes know their locations), n is the path loss exponent, and β is the power absorbed by the receiver serving as the relay node.

Based on the costs, node u selects a route that requires a minimal transfer cost of packets between its neighbors. The chosen node with the minimum cost is the next node to begin the creation of the minimum energy path. Cost measurements are kept up-to-date and broadcast to neighbors. To further optimize energy consumption, the node will switch to sleep mode after completion of the second step if no messages are sent.

To handle dynamic changes in the environment (spread path fluctuations, faulty nodes, etc.), the MECN algorithm also includes a mechanism called "Flip". It is used to deal with the following cases:

1. Nodes are removed from the neighbor set if it is found that direct communication with them is no longer efficient (i.e., it is possible that, due to the dynamic changes in the environment, communication with this node could become more efficient if another neighbor is used as the relay), or
2. A node is added to the neighbor set as direct communication with them becomes more efficient. The newly inserted node activates the cases referred to in point 1.

For the details of the Flip mechanism, Ref. [4] can be referred.

ALGORITHM 1   DISCOVERY OF NEIGHBORS THAT ARE ENERGY-EFFICIENT TO COMMUNICATE

$P_{u \to v \to q}$ is the total transmit power used to transfer packets from node u to q through node v

$N(u)$ is a neighbor set of node u that is energy efficient to connect directly

procedure FINDNEIGHBORSET(u)
  $N(u) \gets \emptyset$
  for all received beacon packets **do**
    $q \gets$ Sender of the beacon
    if $q \notin N(u)$ **then**
      $P_{u \to q}$ ▷ *Compute the power cost*
      N_candidate $\gets$ true
      for all $v \in N(u)$ **and** N_candidate = true **do**
        if $P_{u \to v} + P_{v \to q} < P_{u \to q}$, **then**
          N_candidate $\gets$ false
        end if
        if N_candidate = true, **then**
          $N(u) \gets N(u) \cup \{q\}$
        end if
      end for
    end if
  end for
end procedure

2. **Small Minimum Energy Communication Network (SMECN):** The SMECN [5] algorithm is an expansion of the MECN algorithm. It aims to create a network that is simpler, faster, and more energy-efficient than the one created by MECN [4]. The goal of SMECN is to create a sub-graph G that is smaller than sub-graph G in MECN. As a version of the MECN, the SMECN uses the same energy model and assumptions as the MECN. The implementation of SMECN also consists of two steps identical to MECN. The only difference between SMECN and MECN is the approach used to evaluate the nodes for the enclosure graph. In SMECN, nodes once known to be neighbors are never deleted from the neighbor set and are all included in the enclosure graph. For this cause, SMECN does not need the heuristic "Flip" as in MECN [4]. The work in SMECN has shown that the built subgraph G is smaller than that constructed by MECN while broadcasting at a given power setting is able to enter all the nodes in a circular area around the broadcaster. An energy-efficient reconfiguration algorithm based on SMECN was later represented in Ref. [6].

3. **COMPOW:** The energy-saving approach in COMPOW [7] defines and uses the least standard level of power that is necessary to sustain the connectivity of the whole network. Based on theoretical research, the

authors concluded that the minimum common power level could offer a range of benefits to the networks, including increased transport capability, energy usage, and MAC controversy. The option of using the smallest common power level often results in bi-directional connections, an essential feature needed for efficient routing and proper communication on the MAC layer. This protocol is the first to be implemented in a real wireless testbed and to investigate the different power levels available in the Aironet WLAN access points of the CISCO 350 range. COMPOW blends both power control and routing due to the way that they both influence each other.

COMPOW introduced concurrent routing layer modularity to accomplish the asynchronous and distributed activity. This is achieved by allowing each node to run multiple routing daemons in parallel, one daemon for each transmission power level *Pn*. Thus, each node establishes several routing tables for all available power levels by exchanging hello messages. Initially, each node creates a routing table using the highest power level to locate all the nodes in the network. It then builds a routing table for the remaining power levels and finds the smallest power level whose entries in the routing table are equal to the entries in the routing table at the highest power level. The smallest power level is selected as the optimum power level and its routing table is configured as the master routing table used by the kernel to transfer packets between nodes.

2.2.2 Powder Mode Approach

The power mode approach is a strategy that uses the features of the operating mode available in the network interface of the sensor nodes to conserve resources. There are four modes of operation of the nodes: sleep, idle, send and receive modes. The energy expended during transmission and reception is usually higher than in sleep mode [8]. To send or receive packets, nodes must be in idle mode. However, continuous listening to incoming packets that are not addressed to idle nodes will lead to a high energy dissipation that is very important compared to sleep mode [3]. This implies that redundant nodes sitting in the idle can be switched to the energy-saving mode by having them in sleep mode. This function has been used in topology control to conserve resources and extend network life without losing network power and communication. In this section, three power mode algorithms dealing with power-off idle nodes as well as coordinating sleep and wake-up scheduling of nodes are discussed.

1. **Geographical Adaptive Fidelity (GAF):** The core concepts of GAF [9] are to provide an appropriate number of nodes that remain active for communication and to put redundant nodes in sleep mode without disrupting network connectivity. To distinguish active nodes from redundant nodes, GAF splits the network area into small virtual grids. Both nodes are connected to these grids by using position information and an idealized radio model. Figure 2.6 illustrates an example of three simulated grids.

IoT and Topology Control

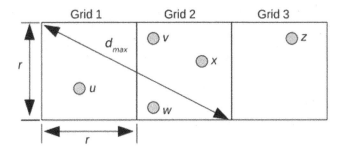

FIGURE 2.6 The virtual grid structure in GAF.

The length of the grid r is selected in such a way that every two nodes in the adjacent grids will meet each other. The scale of the simulated grids is based on the nominal R spectrum and they are all equal in size. The longest potential distance between the nodes of the adjacent grids is the length of the diagonal connecting the two adjacent grids that can be measured using Equation (2.2),

$$r \leq R/5 \qquad (2.2)$$

GAF uses the word "equivalent nodes" to describe a neighbor set appropriate for routing. The "equivalent nodes" identify nodes capable of interacting with all the nodes in their neighboring grids. Equivalent nodes can be used to save power by keeping only some of them alive for routing while the remaining nodes can sleep. For example, in Figure 2.6 [9], nodes v, w, and x are identical nodes, since, in order for node u to communicate with node z, packets can be relayed by either v, w, or x. In this example, energy savings are accomplished by putting nodes w and x in sleep mode while node v forwards data and alternates between sleeping and active. This phase can be represented in a state transformation diagram in Figure 2.7, which is redrawn from Ref. [9].

There are three GAF operational states – primarily sleeping, discovery, and active states. In the discovery state, nodes identify their neighbors on the grid by flipping on their radios and exchanging discovery messages. The discovery message contains the id of the node, the state, the grid, and the active time of the node T_{act}. The T_{act} value is used to evaluate the length of the nearby nodes remaining to sleep. The instant when the node depletes half of its energy resource is set. Nodes engage in routing in the active state. In the sleeping state, the nodes turn off the radio and stay inactive. Initially, all nodes begin with the state of discovery. In this state, nodes set their discovery time for T_d seconds, transmit the discovery message to find nodes inside the same grid, and then enter the active state. Nodes that join the active state set their timer to the timeout value T_a to determine the period that they remain in the active state. After T_a, the nodes will return to the discovery state and rebroadcast their discovery message every T_d second. Nodes in discovery or active state can turn to sleep state if they find other equivalent nodes for routing. When going to sleep, nodes cancel all the timers and shut down their radios. They sleep for the length of T_s, which is a random interval between $T_{act}/2$ and T_{act}. Information

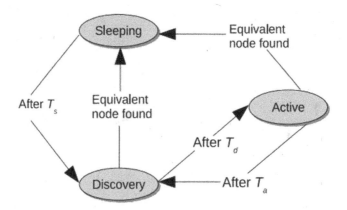

FIGURE 2.7 State transitions in GAF.

on whether or not these nodes are capable of receiving messages during sleep is not given in Ref. [9].

To optimize network life, nodes that engage in routing are ranked according to a set of rules. The rules ensure that only one active node remains in each grid and that nodes with a longer projected lifespan are used first. There are a variety of rules for determining rank. Next, the node in the active state is of a higher rank than the node in the discovery state. Second, if nodes are in the same state, the GAF gives a higher rank to a node with a longer life expected. Third, node IDs are used to remove the connection. GAF also adopts a load balancing approach to ensure that the load is equally spread between nodes to keep nodes from exhausting their energy. By setting the timeout value T_a, nodes that are in an active state can gradually move to the exploration state to allow other nodes with a higher energy level to become active within the same grid. GAF considers system-level actions to conform to high mobility to avoid the drop of high packet rates. This is achievable by calculating the time that each node persists in the grid. This value of the T_{mob} is included in the discovery message and forwarded to the neighbors. Its neighbors who are about to transit to sleeping state use the T_{mob} along with the T_{act} to assess the length of their T_s.

2. **Sparse Topology and Energy Management (STEM):** The concept of STEM [10] is to position as many nodes as possible in sleep mode in such a way that energy consumption is minimized and the network life is increased. The system proposed that this idea is relevant to a network that spends much of its time monitoring operations and has less data forwarding activity. The inactive nodes that track the operation can only be shut down and woken up when they have data to be sent to the base station. The common challenge of the power-down approach is to control the sleeping transfer of nodes so that sleeping nodes are triggered only when an incident occurs. STEM addresses this problem by changing the node's radio on regularly for a brief time to listen to incoming communications.

IoT and Topology Control

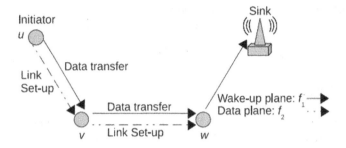

FIGURE 2.8 Radio setup of a sensor node in STEM.

There are two activities involved in STEM, the wake-up and data transfer systems. The wake-up process ensures that the sleeping node radio is switched on to allow nodes to respond to the incoming call, while the data transmitting process ensures that data is securely transmitted between the source and the sink. In STEM, each node sends a wake-up message and transmits data to two different frequency bands using two separate radios to prevent interference. The wake-up alert occurs on the wake-up plane operating on the radio frequency f_1 while the data transmission occurs on the data plane operating on the radio frequency f_2. The STEM operation is seen in Figure 2.8.

Considering that node v and w are asleep, assume that node u senses an occurrence and wants to send data to the sink by node v and w. Node u then sends a wake-up message to the target node v on the radio frequency f_1 and waits for the node v response. After receiving the response, both nodes turn on their radios and start transmitting data to the radio frequency f_2. This process is replicated between node v and node w, with node v being an initiator, while node w becomes a target before the data is successfully retrieved by the sink.

Later development combines STEM with GAF in order to accomplish two goals. The first goal is to allow further energy saves. GAF means that each grid must have one active node operating as a leader, but that leader does not have the data to be transmitted. Thus, by running STEM on the leader in each GAF grid, the leader who is sitting idle waiting for data transmission may be shut off to minimize power consumption. The result shows that the interconnected STEM decreases energy consumption by up to 7% relative to a network without any topology power. This change is equal to an increase of 14% in the lifetime of the node as stated in [10]. The second goal is to increase the latency of STEM. STEM uses the GAF leader election mechanism to minimize the amount of intrusion during the wake-up process and speed up the connection set-up step.

3. **Adaptive Self-Configuring Sensor Network Topologies (ASCENT):** ASCENT [11] is a self-configured algorithm that allows nodes to calculate operating conditions locally. Based on these conditions, the nodes then determine whether or not they need to engage in the routing. To achieve energy conservation, ASCENT chooses a subset of nodes to stay involved to serve as a routing backbone. The remaining nodes in the network stay

inactive, listen to other nodes, and search regularly if they need to join the routing backbone. For example, when the packet loss is high, passive nodes are enabled to retain communication. Otherwise, these nodes would turn off their radio to conserve energy.

Nodes in ASCENT will remain in one of four states, namely test, passive, active and sleep, as seen in Figure 2.9.

Nodes in passive state remain in listening mode to engage in routing if necessary, so the node radio remains active. Nodes in active state conduct data transfer and control while nodes in sleep state turn off their radios to conserve energy. Nodes in the test state verify whether the network has sufficiently active nodes to ensure connectivity. If there are insufficient active nodes, the nodes in the passive state will either join the active nodes or move to the sleeping state to minimize energy usage.

Initially, all nodes stay in the test state. The nodes then set their timer T_t and send messages to the neighbors to discover their neighbors. When in the test state, the node will verify if the number of active neighbors N is above the neighbor NT threshold or the average data loss rate DL is greater than the average loss T_o before entering the test state. If the condition is valid, the node is shifted to the passive state. The higher node ID in the announcement message is used to break the tie if several nodes struggle for transport to the test state. If the condition is incorrect, the node stays in the passive state and switches to the active state after the timer T_t has elapsed. In the active state, the node engages in the routing until it has run out of energy. The active node sends support messages when the DL is higher than the LT failure threshold.

A node that joins the passive state sets up a T_p timer. It sends new passive node announcement messages to be used by active nodes to measure the overall density of the nodes in the neighborhood. When in a passive state, the node chooses whether to travel to the test state to help the routing spine or transit to the sleeping state to conserve energy. The decision to transit to this state shall be taken locally if any of the two conditions are met: 1) The number of neighbors is below NT and DL

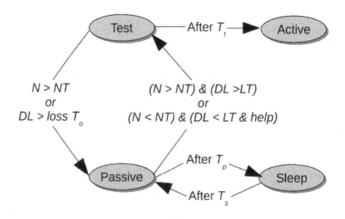

FIGURE 2.9 State transitions representing the operation of ASCENT.

is greater than *LT*. 2) The number of neighbors is below *NT*, *DL* is below *LT*, and the node gets support from active neighbors. Otherwise, the timer T_p will remain in a passive state until it expires. Later, the node moves to sleep and turns off its radio to save energy. Upon expiry of the timer T_s, it will be moved to the passive state.

In the case of ASCENT, the neighbor threshold *NT* and the failure threshold *LT* values may be set to satisfy the requirements of the applications. *NT* can be modified to increase network performance, and *LT* can be picked to reduce packet loss. Other parameters, such as T_t, T_p, and T_s, are calculated dynamically over time to increase power efficiency but trade-off connection consistency.

2.2.3 Clustering Approach

The principle of clustering is to pick a collection of nodes on the network to create an efficient topology. The selection of neighbors can be made on the basis of different parameters, namely energy reserve, network density, or node identifier. Unlike the power adjustment or power mode approach, the clustering approach builds a topology of hierarchical structures that are scalable and easy to handle. The benefit of clustering is that a certain role can be confined to a group of nodes called cluster heads and can be allocated to receive, process, and forward packets from nonclusterheads. This process allows for an effective organization of the network. Other appealing features of clustering methods include load balancing and data aggregation or data compression provided for extended network life. The selection of clusterheads remains fixed in certain clustering approaches. As a result, cluster heads usually undergo faster energy depletion because they are heavily loaded with different tasks. This problem is solved by randomizing the selection of clusterheads to distribute loads equally among the nodes on the network.

Many clustering approaches create a virtual backbone using the connected dominant set (CDS) definition. CDS has been commonly used as a topology monitor for the conservation of network energy resources. The dominant set (DS) is defined as a subset of nodes in the graph so that each node not in the subset has at least one direct node belonging to the subset [12]. If the nodes in the dominant set form a linked graph, the set is called a CDS. Figure 2.10 shows an example of a CDS created in a network consisting of 14 nodes. In this figure, nodes *u*, *v*, *w*, *x*, *y*, and *z* form the backbone for data forwarding while the remaining nodes do not participate in data forwarding. This technique eliminates overhead and energy communication. The following section addresses the three clustering algorithms used to control topology.

1. **Power Aware Connected Dominating Set (PACDS):** Wu et al. [13] suggested a simple algorithm based on the CDS principle that would find a CDS using a simple marking method. PACDS strengthens the work proposed in Ref. [14] to accomplish two objectives. The first goal is to create a small CDS while the second goal is to extend the life of the nodes. In the CDS, nodes in the backbone are usually overwhelmed with a variety of tasks and are the first in the network to drain energy. Load balancing can overcome this problem by randomizing the role of the backbone between nodes with higher residual energy.

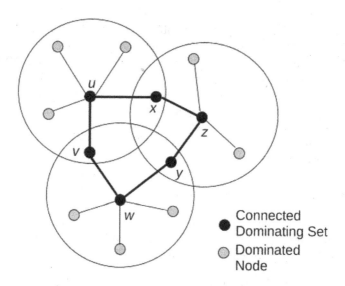

FIGURE 2.10 A backbone in the network is built using a connected dominating set.

The construction of the PACDS consists of two stages, as shown in Algorithm 2. The first stage is the development of a CDS. Initially, node u broadcasts a hello message to collect information from the neighbor. If node u has two unconnected neighbors, it will be named as a CDS node. The second stage is the pruning process in which redundant CDS nodes are removed to reduce the size of the CDS. The process of pruning is necessary because the size of the CDS formed during the first stage is not minimal. Two rules that are based on the node ID are used to remove the CDS. The pruning rules state that if node u has a neighbor with a higher ID that can cover all of its neighbors or if u has two neighbors with a higher ID that can cover all of its neighbors, u can be eliminated from the CDS.

Apart from the use of node ID, PACDS implements additional CDS pruning rules using node and residual energy. The first additional pruning rule uses node degree intending to keep the size of the CDS as minimal as possible. The second additional rule uses residual energy to achieve an extended life span of the node. The residual energy rule gives higher priority to nodes with a higher energy level to become cluster heads and excludes the lower energy level nodes from the CDS.

2. **Energy-Efficient Distributed Connecting Dominating Sets (ECDS):** Yuanyuan et al. [15] proposed ECDS to resolve energy constraints in wireless sensor networks and to minimize the size of the CDS. ECDS is a topology control algorithm based on CDS. The features of ECDS are the enhancement of the network life and equal distribution of energy achieved by load balancing. In comparison to PACDS [13], ECDS first constructs a dominant set that is a maximum independent set (MIS) and then seeks gateway nodes to connect to the MIS. In addition, ECDS does not use any pruning mechanism as in PACDS [13]. An MIS is classified as an independent set (*IS*), which is not a subset of any other *IS* [16]. The IS of

IoT and Topology Control

ALGORITHM 2   CONSTRUCTION OF A CDS IN PACDS ALGORITHM

Require: Graph G(V,E)
Ensure: Connected Dominating Set *(CDS)* ∈ G(V, E)
procedure FORMCDS*(G(V,E))*   ▷ Create a CDS
   CDS ← ∅
  for all $u \in V$ **do**
    for all $v, w \in N(u)$ **do**
     if $w \notin N(v)$, **then**   ▷ unconnected neighbors
      CDS ← CDS ∪ {u}
     end if
    end for
  end for
end procedure
procedure PRUNECDS*(G(V,E))*   ▷ Remove redundant CDS needs
  for all $u \in CDS$ **do**
   if u satisfies the pruning rules, **then**
    CDS ← CDS − {u}
   end if
  end for
end procedure
procedure PACDS*(G(V,E))*   ▷ Main procedure
  FORMCDS*(G(V,E))*   ▷ Phase 1 of the CDS formation
  PRUNECDS*(CDS)*   ▷ Phase 2 of the CDS formation
end procedure

Graph G is a subset of V where there are no two nodes inside the set of the edge. Notation V refers to a set of vertexes V in graph G. Therefore, every MIS is a dominant set that is not connected.

The construction of the ECDS consists of two steps, as demonstrated in Algorithm 3. The first phase determines the MIS using a coloring technique to distinguish MIS nodes from non-MIS nodes while the second phase selects the connectors to join the MIS. The processes involved are seen in Algorithm 4. Initially, all nodes are in white and at the end of the first step, nodes will be either in black (MIS nodes) or grey (non-MIS nodes). The first step begins with the initiator volunteering to be the MIS node and coloring itself black. It sends a black message to its neighbors. The white neighbors who got the message are colored grey (non-MIS nodes) and send a grey message to their white neighbors to update their color changes. The white nodes that obtain the grey message are not MIS nodes, but they are possible MIS nodes. These nodes send an inquiry message to their neighbors to know their states and weights, and to wait for their responses for the duration of the timeout. The white nodes with the highest weight are selected as MIS nodes. The weight is

---

ALGORITHM 3 CONSTRUCTION OF A CDS IN ECDS ALGORITHM

Require: Graph $G(V,E)$ and initiator node i
Ensure: Connected Dominating Set $(CDS) \in G(V,E)$
    procedure ECDS$(G(V,E))$ > Main procedure
        GENERATEMIS$(G(V,E))$ > Phase 1 of ECDS
        FINDCONNECTOR$(G(V,E))$ > Phase 2 of ECDS
    end procedure

---

determined on the basis of residual energy (E_{res}) of the node and the degree of efficiency (D_{eff}). The effective level is determined by the number of neighbors in the white and transition states. When they receive a black message during the timeout, they become grey nodes (non-MIS nodes). In other words, they are MIS neighbors and cannot be MIS nodes. Otherwise, they will remain in the transition state until the timeout is over. After the timeout, if these nodes are of the greatest weight among all neighbors, they become a black node and the process is repeated.

The second step determines the MIS gateway nodes and connects the MIS to create the CDS as shown in the Algorithm 4.

These gateways are not MIS members. System runs a greedy algorithm to locate connectors, where each MIS node selects the highest-weight non-MIS node that can connect to other MIS nodes within two-hop communication. After completion of the greedy algorithm, all nodes in the network are either blue (CDS nodes) or grey (non-DS nodes) in color. The second step begins with MIS nodes that send a call message to non-MIS (grey) nodes that have the ability to be connectors. Upon receipt of the request message, the grey node calculates its weight and sends an update message to the MIS nodes. The weight is determined in the same way as in the first step, except that the effective degree is specified by the number of nodes in the MIS and the black states. Based on the weight determined, the MIS nodes choose the most weighted grey nodes to be connectors. The CDS construction stops if any MIS blue colored node terminates the algorithm or terminates any non-MIS node that satisfies the following conditions: (i) it is blue colored (connector node) or (ii) all its neighbors are blue and grey colored.

3. **Topology Management by Priority Ordering (TMPO):** TMPO [17] is a dynamic algorithm that considers movement and residual energy when the backbone is formed. TMPO introduces the idea of gateways and gateways (which are used in clustering methods) to connect dominant sets. Several outstanding features of the TMPO are summarized in Ref. [17]. First, the creation of minimal dominant sets and the CDS is free from any negotiation process, thereby preventing excessive overheads during the Cluster head election. Second, the identifier called the node priority is determined regularly, allowing the cluster head to rotate between nodes to extend the existence of the network. Third, when choosing cluster heads, the

IoT and Topology Control

ALGORITHM 4 PHASE 1 AND PHASE 2 OF THE ECDS ALGORITHM

procedure GENERATEMIS$(G(V,E))$ ▷ Phase 1: Find a MIS in the network
    $MIS \leftarrow \emptyset$; $GRAY \leftarrow \emptyset$
    $MIS \leftarrow MIS \cup \{i\}$
    for all $u \in N(i)$ **do**
      $GRAY \leftarrow GRAY \cup \{u\}$
      for all $v, w \in N(u)$ **do**
        $weight(v) \leftarrow E_{res}(v) \times D_{\textit{eff}}(v)$
        $weight(w) \leftarrow E_{res}(w) \times D_{\textit{eff}}(w)$
        if $weight(v) > weight(w)$, **then**
          $MIS \leftarrow MIS \cup \{v\}$ ▷ Black nodes
        else
          $GRAY \leftarrow GRAY \cup \{v\}$ ▷ Gray nodes
        end if
      end for
    end for
end procedure

procedure FINDCONNECTOR$(G(V,E))$ ▷ Phase 2: Find connectors to join the MIS
    $CDS \leftarrow \emptyset$; $CONNECTOR \leftarrow \emptyset$
    $CDS \leftarrow CDS \cup \{i\}$ ▷ Initiator invokes the connector process
    for all $u \in N(i)$ **do**
      if $u \in CONNECTOR$, **then**   ▷ Set of nominated connectors
        $CDS \leftarrow CDS \cup \{u\}$ ▷ Blue nodes
      end if
    end for
    for all $v \in MIS$ **do**
      if v has $x \in CDS$, **then**
        $CDS \leftarrow CDS \cup \{v\}$ ▷ Blue nodes
      else
      for $w \in N(v)$ **do**
        $weight(w) \leftarrow E_{res}(w) \times D_{\textit{eff}}(w)$
        if w has $max(weight)$, **then**
          $CONNECTOR \leftarrow CONNECTOR \cup \{w\}$
        end if
      end for
      end if
    end for
end procedure

algorithm considers the mobility and energy power of nodes. Fourth, unlike other clustering approaches that use gateways and cluster heads to form a cluster, TPMO introduces a new concept called doorway.

There are two phases involved in the construction of a CDS as shown in Algorithm 5.

The first step is the selection process for the cluster. Finding cluster heads that can create a minimum dominant set in the network to minimize the size of the CDS is done. The selection of cluster heads shall be made in compliance with the priority rule. A node becomes a cluster head if it has the highest priority among its one-hop neighbors or the one-hop neighbors of one of its one-hop neighborhoods. The priority between the candidate cluster head nodes is the identification of the neighbor node, the present time, and the importance of the willingness. The value of willingness is assigned to each node u as a function of node mobility (s) and energy level (E_u). It shall be measured as Equation (2.3),

$$W_u = 2^{\log 2 (0.9 E_u) \log 2 (s i \mid 2)} \tag{2.3}$$

Based on the value of the willingness, the node identifier, and the current time, the priority of the node is determined. The priority values of the node u ($u_{priority}$) are updated regularly to provide a random set of cluster heads, and these values are special. It shall be determined by Equation (2.4),

$$u_{priority} = \mathbf{Hash}\left(\left(t - u_{off}/T\right) \oplus u\right) \times W_u \oplus u \tag{2.4}$$

where Hash represents a pseudo-random number generated in the range of 0 to 1, u_{off} is the time slot of the node u, and $\oplus$ is the bit-concatenation operation.

In the second phase, the doorway and gateway nodes are selected and they connect the minimal dominant set created in the previous phase to form the CDS. A doorway node is defined as a node that can connect two cluster heads that are separated from each other by three hops and there are no other cluster heads between them. The doorway must have the highest priority between the two clusters. The gateway node is known as the highest priority node that can link two cluster heads to two hop-offs or connect one cluster head and one gateway separated two hop-offs and there are no other cluster heads between them. After the choice of gateway and gateway nodes, the CDS is created.

2.2.4 Hybrid Approach

The hybrid approach is a topology control technique that uses some form of clustering in combination with other approaches such as power adjustment or power mode to achieve additional energy savings. The following section introduces three hybrid algorithms that are designed to conserve energy.

ALGORITHM 5 CONSTRUCTION OF A CDS IN TMPO ALGORITHM

Require: Graph $G(V,E)$
Require: a set of relay nodes with the highest priority between two clusterheads that are three-hop away *(3HRELAYSET)*
Require: a set of relay nodes with the highest priority between two clusterheads that are two-hop away *(2HRELAYSET)*
Ensure: Connected Dominating Set *(CDS)* $\in G(V,E)$

procedure ELECTCLUSTERHEAD$(G(V,E))$ ▷ Select Clusterheads
    $CLUSTERHEAD \leftarrow \emptyset$; $CDS \leftarrow \emptyset$
    $u_{priority} \leftarrow \mathbf{Hash}((t-u_{off}/T) \oplus u) \times W_u \oplus u$
    for all $v \in N_1(u)$ **and** $w \in N_2(u)$ **do**
        $v_{priority} \leftarrow \mathbf{Hash}((t-v_{off}/T) \oplus v) \times W_v \oplus v$ ▷ *one-hop*
        $w_{priority} \leftarrow \mathbf{Hash}((t-w_{off}/T) \oplus w) \times W_w \oplus w$ ▷ *two-hop*
    end for
    if $(u_{priority} > v_{priority})$ *or* $(u_{priority} > w_{priority})$, **then**
        $CLUSTERHEAD \leftarrow CLUSTERHEAD \cup \{u\}$
        $CDS \leftarrow CDS \cup \{u\}$
    end if
end procedure

procedure FORMCDS$(G(V,E))$ ▷ Form a CDS
    for $u, v \in CLUSTERHEAD$ **do**
        if $3HRELAYSET \cap CLUSTERHEAD = 0$, **then**
            for $x \in 3HRELAYSET$ **do**
                $CDS \leftarrow CDS \cup \{x\}$ ▷ Doorway node
            end for
        end if
        if $2HRELAYSET \cap CLUSTERHEAD = 0$, **then**
            for $z \in 2HRELAYSET$ **do**
                $CDS \leftarrow CDS \cup \{z\}$ ▷ Gateway node
            end for
        end if
    end for
end procedure

procedure TMPO$(G(V,E))$ ▷ Main procedure
    ELECTCLUSTERHEAD$(G(V,E))$ ▷ Phase 1 of TMPO
    FORMCDS$(G(V,E))$ ▷ Phase 2 of TMPO
end procedure

1. **SPAN:** The SPAN [18] algorithm uses a hybrid power mode and clustering approaches. Selects a subset of nodes to form a forwarding backbone using a CDS approach. The backbone is capable of forwarding packets, maintaining network access, and maintaining network power. Based on local decisions, SPAN nodes determine whether to enter or sleep in the forwarding backbone. Nodes in the forwarding backbone are called coordinating nodes, while the remaining nodes in the network are called non-coordinator nodes. Non-coordinator nodes stay in sleep mode to conserve power and regularly wake up to share traffic with coordinating nodes. One of the key advantages of SPAN is the use of the 802.11 power-saving features to increase routing throughput and packet transmission latency. Using SPAN at the top of the 802.11 power-saving modes, packets sent to sleep mode can be deposited temporarily in the neighbor. The packets are later recovered as the node wakes up, avoiding the failure of the packet.

SPAN is structured to achieve the following four objectives [18]. Second, it elects an appropriate number of coordinators to ensure that each node has at least one coordinator from its one-hop neighbors to ensure the network compatibility. Second, it uses a load balancing strategy that rotates the coordinators to ensure that the organizing role is equally dispersed across all nodes. Second, it produces a limited size of CDS. The produced CDS may not, however, be minimal. Fourth, the identification of coordinators is rendered in a clustered manner using dispersed information obtained from neighbors.

The activity of the SPAN is controlled by two mechanisms called the election coordinator and the withdrawal of the coordinator. These two processes are seen in Figure 2.11. The information needed to withdraw or elect a node as coordinator is shared between neighbors through HELLO messages. During the coordinating election, the non-coordinator node regularly reviews if it is chosen as the co-ordinator on the basis of the eligibility requirement for the coordinator. The rule

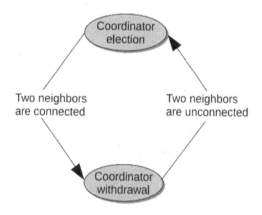

FIGURE 2.11 The state diagram of the coordinator election and withdrawal processes in SPAN.

IoT and Topology Control

notes that a non-coordinator node would become a coordinator if it has two neighbors who cannot connect directly or via intermediate coordinators. In the case of a conflict that arises when several nodes attempt to become coordinators at the same time, SPAN uses a random back-off delay to settle the dispute. This is achieved by setting a random delay and reporting the conflict to the neighbors. When the wait has elapsed, the nodes will re-check the coordinator's election by forwarding the coordinator's announcement. If the qualifying requirement is still valid, they can become coordinators. Interestingly, SPAN combines both the energy level of the nodes and the capacity of the nodes to link additional pairs of nodes between their neighbors to resolve the problem of energy conservation and equal allocation of loads in the network.

The coordinator withdrawal process enables the role of coordinator to be shared between nodes. When a co-ordinator has both neighbors that may connect individually or via other co-ordinators, it withdraws from becoming a co-ordinator and becomes an ongoing co-ordinator. It stays in that state for a certain amount of time until WT withdraws its position as coordinator. SPAN findings show that it can sustain network compatibility, maintain capacity, and have substantial energy savings. SPAN simulations demonstrate that machine life with SPAN is more than a factor of two better than without SPAN.

2. **CLUSTERPOW:** The CLUSTERPOW [19] algorithm blends a clustering approach with a power control approach to network access, network capacity, and energy efficiency. The architecture of CLUSTERPOW is inspired by the weakness of COMPOW [7] in the handling of non-homogeneous node distributions. The alternative of using the minimum common power level in COMPOW is not suitable for non-homogeneous networks since the lowest common power level is determined by a distant node. For example, see Node u in Figure 2.12 redrawn from [19]. All the nodes within the *C1* cluster use 1mW of power to communicate. When a node w joins the network, the majority of the nodes in cluster *C1* are required to use an excessively higher power level of 100mW to connect with node w. As a consequence, the minimum common level of power is set at a much higher level. As a solution, CLUSTERPOW provides a common topology control and routing solution that selects an optimal minimum power level for each

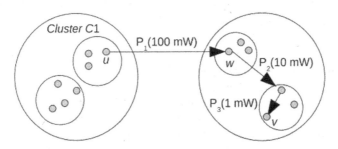

FIGURE 2.12 The CLUSTERPOW multihop routing using a smaller power creates a non-homogeneous network.

cluster. CLUSTERPOW offers implicit clustering, which ensures that the selected limited transmit power level instantly generates clusters. As a consequence, it has no cluster heads or gateways.

In the same way as COMPOW, CLUSTERPOW allows each node to hold separate routing tables, one for each power level built using HELLO messages exchanged. It also uses parallel modularity on the network layer by running several routing daemons as in COMPOW. When node u has a message to send to node v, it determines the lowest transmission power level P such that the destination can be reached in several hops by using a power level lower than P. This method is carried out at the source and every intermediate node along the path from the source to the destination. For example, the network in Figure 2.12 has three clustering levels corresponding to the 1mW, 10mW, and 100mW power levels. To relay from node u to node v, a power level of 100mW is used at each hop before the packet enters the 10mW cluster. Then 10mW is used at each hop and the transmission power is reduced to 1mW as the packet gets closer to the destination.

3. **Low-Energy Adaptive Clustering Hierarchy (LEACH):** LEACH was implemented in [20] to minimize energy consumption by clustering, data aggregation, load balancing and TDMA/CDMA. LEACH combines a clustering approach and a power mode approach to extend the life of the network. A considerable amount of energy is used in WSNs as both nodes engage in long-distance data transfer. Using clustering will minimize the energy spent by limiting the number of nodes participating in long-distance transmissions. In clustering, only cluster-headed nodes can send data to the base station. LEACH assigns data aggregation and fusion tasks to cluster head nodes to minimize the number of data transmissions. LEACH uses a load balancing system that rotates the position of cluster head nodes periodically. An equal and clear collection of cluster head nodes is often used to ensure that nodes die at random. LEACH uses TDMA to eliminate intra-cluster interactions that address collisions, hidden conflicts, overhearing, and idle listening.

This is done by shutting off non-cluster head node radios while they are not in service. LEACH uses CDMA to solve collisions between cluster head nodes vying for simultaneous data transfer to base stations.

The LEACH operation is split into rounds with each round consisting of two stages. The two steps are seen in Algorithm 6. The set-up process is responsible for cluster creation, while the steady-state phase is responsible for forwarding data to the base station. The set-up process starts with the selection of the cluster head. Cluster head options are rotated in each round to have a consistent load distribution and prolong the life of the node. Cluster heads are selected randomly on the basis of two parameters. The parameters are based on the proposed percentage of cluster heads and the number of cluster heads of the node. Each node u chooses a random number, with an RN between 0 and 1. If the number is less than the threshold value, then Th becomes the cluster head for the round. The Th is being measured as Equation (2.5),

IoT and Topology Control

$$Th(u) = \begin{cases} \frac{p}{1-p(r*1/p)}, & \text{if } n \in G \\ 0, & \text{Otherwise} \end{cases} \quad (2.5)$$

If P is the optimal percentage of the cluster head, 'r' is the current round, and 'G' is the non-cluster head collection in the last 1/P rounds. Therefore, the chances of a node being a cluster header are poor if it has been chosen as a cluster header in the previous round. The nodes of the selected cluster then broadcast their election to the rest of the nodes on the network by sending an INVITE message. The non-cluster head nodes that receive the INVITE message would then calculate the signal intensity obtained to select a cluster. The nodes will join the cluster with the highest signal intensity value and notify the cluster nodes of their decision. This information is needed by the cluster head nodes to create a TDMA schedule for each member of the cluster.

In the steady-state process, nodes that are scheduled for data transmission will begin the transmission of their data to the cluster head node. Nodes that are not planned for transmission will switch to sleep mode to maintain power. The data collected by the cluster head nodes are aggregated or combined to compress the size before being sent to the base station. For some time, the next round will commence again and the two stages will be repeated.

2.3 COMPARATIVE ANALYSIS: TOPOLOGY CONTROL METHODS

So far, we've addressed and categorized topology control algorithms into four groups. For a deeper understanding of the success of various algorithms from an energy consumption perspective, have included a cost analysis depending on the implementation parameters. Direct comparisons between algorithms are not feasible, because various parameters are used to measure the energy efficiency value of algorithms, thereby posing a challenge in finding common ground for evaluating algorithms. It should be noted that the objective comparison of these algorithms is not feasible due to the lack of information as per the context. The cost comparison of their characteristics is summarized in Table 2.1.

In Table 2.1, N represents the total number of nodes, V is the number of neighbors, P is the number of transmission power levels, and 'Δ' is the highest degree in the graph.

A discussion on each algorithm in the four categories is given in the following section. In order to provide an overall comparison of these algorithms, Table 2.1 listed the benefits and drawbacks of these algorithms. In addition, the analysis of the network existence description used by each algorithm demonstrated the advantages and disadvantages of these concepts.

1. **Power Adjustment Approach:** This section presents the topic of power adjustment algorithms, discussing their advantages and drawbacks, as well as the network existence concept used.

The good connectivity of the enclosure graph [4] is one of the main results of the Minimum Energy Communication Network (MECN). In the worst case, each node is capable of establishing communication relations with all the nodes inside its

TABLE 2.1
Cost comparison of distributed topology control algorithms

Algorithms	Localized	Time complexity	Message complexity	Space complexity	Connectivity	Mobility
MECN	Yes	$O(v^3)$	Not provided	$O(V^2)$	High	Low
SMECN	Yes	$O(v^2)$	Not provided	Not provided	High	Low
COMPOW	No	Not provided	$O(Pn)$	Not provided	Low	Low
GAF	No	Not provided	$O(V)$	Not provided	Low	Low
STEM	Yes	Not provided	Not provided	Not provided	Low	No
ASCENT	Yes	Not provided	Not provided	Not provided	Low	No
PACDS	Yes	$O(\Delta^2)$	$O(n\Delta)$	Not provided	High	Low
ECDS	Yes	$O(n)$	$O(n)$	Not provided	High	No
TMPO	Yes	Not provided	Not provided	Not provided	High	High
SPAN	Yes	$O(n)$	$O(n)$	Not provided	Low	Low
CLUSTERPOW	No	Not provided	$O(Pn)$	Not provided	Low	Low
LEACH	Yes	$O(n)$	$O(n)$	Not provided	Low	Low

enclosure [4]. MECN also creates a sparse network, which means that the number of connections grows linearly with the number of nodes on the network. The effect of the sparse network is to reduce the degree of disturbance and enhance energy conservation. There are, however, some predictions made in the MECN algorithm. The presumption that all nodes know their precise positions in the deployment area via the Global Positioning System (GPS) is not realistic. This is due to the message overhead created for updating location information and also for installing additional hardware.

MECN also believes that each node can connect with all its neighbors and neglects the barriers that normally occur between two nodes in the deployment area. Another drawback of MECN is its dependency on an explicit propagation model for the computation of the relay area and the enclosure graph [21]. For example, in order to decide the lowest energy path, nodes need to measure all feasible routes depending on the actual level of transmission capacity. Therefore, when calculating the optimal topology, realistic radio transmission conditions must be used.

One of the key problems mentioned in [4] is the restriction of the search area to the end of the algorithm. When the nodes are extremely mobile, the estimation of the relay nodes and the enclosure area can be expensive (time complexity) and the energy of the nodes can be depleted. The time complexity of the MECN given in [22] is $O(V^3)$, where V is the number of node neighbors. Although the message

---

ALGORITHM 6   THE LEACH ALGORITHM

Require: Graph $G(V,E)$
Ensure: Cluster $(C) \in G(V,E)$
procedure $SETUPCLUSTER(G(V,E))$ ▷ Form clusters
    $C \leftarrow \emptyset$; $CH(u) \leftarrow \emptyset$
    $RXSS \leftarrow \emptyset$ ▷ Store the received signal strength of clusterheads
    for all $u \in V$ **do**
      $RN(u) \leftarrow RN[0:1]$ ▷ Choose a random number
      if $RN(u) < Th(u)$ **then**
        $C \leftarrow C \cup \{u\}$
        $CH(u) \leftarrow CH(u) \cup \{u\}$ ▷ Selected as a clusterhead
      end if
    end for
  for all $v \in N(CH)$ **do**
    $RXSS(v) \leftarrow RXSS(v) \cup \{SSI(CH)\}$ ▷ SSI, received signal strength of clusterhead
    for all $SSI(w), SSI(x) \in RXSS(v)$ **do**
    if $SSI(w) > SSI(x)$ **then**
      $CH(v) \leftarrow CH(v) \cup \{w\}$
    else
      $CH(v) \leftarrow CH(v) \cup \{x\}$
    end if
    end for
  end for
end procedure

procedure $STEADYSTATE(G(V,E))$
    if u scheduled for transmission **then**
      transmit and aggregate data
    else
      transit to sleep mode
    end if
end procedure
procedure $LEACH(G(V,E))$ ▷ Main procedure
    $SETUPCLUSTER(G(V,E))$ ▷ Phase 1 of LEACH
    $STEADYSTATE(G(V,E))$ ▷ Phase 2 of LEACH
end procedure

---

overhead of MECN is not given, we assume that a substantial overhead message is also added during the second step, in which MECN relies on global knowledge to compute the best topology. MECN is optimized for static or slow-moving networks. However, due to its localized property, it is also suitable for mobile networks, though likely at the expense of a relatively high overhead message.

The Small Minimum Energy Communication Network (SMECN) uses the same network and energy models as the MECN [4] network. It also provides both the advantages and drawbacks of MECN for common functionality. SMECN outperforms MECN in terms of power efficiency and time efficiency due to its lower generated subgraph. The complexity of the time of SMECN is O(V2) [22]. It converges faster than MECN because the subgraph built has fewer connections, which can also result in lower maintenance costs for the connection and achieve substantial energy savings [5]. It is noted that during a neighbor search, the option of transmission power is determined by the density of the network. For example, lower transmission power is necessary to enclose a dense network, whereas a much higher transmission power is required to enclose a sparse network. This indicates that SMECN might not be a power-efficient approach for sparse networks where the highest transmission power is commonly used. In such situations, the node battery will easily drain and shorten the life of the network. This restriction was discussed in [22].

COMPOW has a modular structure that allows the control of topology to be implemented into every constructive routing protocol, making it versatile. There are, however, some deficiencies of COMPOW. The first shortcoming is its significant overhead message. Each node has six separate power levels and shares substantial state information with other nodes to set the optimal power level. This method generates an additional message overhead, which can drain the energy reserve of the nodes and shorten the life of the nodes. The decision to set the optimum power level is often taken on the basis of global details given by different routing tables running several power levels.

As a result, there is a significant overhead message to manage and refresh the network topology. In the worst-case example, the overhead message of COMPOW is $O(Pn)$ [23], where P is the number of power levels used by nodes, and n is the total number of nodes in the network. If nodes run more power levels, a large message overhead is expected. In realistic circumstances, P can be as high as 10 as stated in [24]. The second shortcoming of COMPOW is evident in a non-homogeneous network where nodes are expected to converge to a far higher common level of power set by a new node entering a network that is far removed from the majority of nodes. As a result, higher transmission power is used to sustain the network graph, thereby defeating the goal of reducing power usage by using a minimum common power level.

2.3.1 Evaluations Based on the Network Lifetime Definitions

Among the three power adjustment algorithms, only SMECN gives a network life definition, while MECN and COMPOW do not define a network life definition.

The goal of MECN is to optimize the battery life of the network by discovering the most energy-efficient routes. Depending on the accuracy of the radio model, the overall energy usage for the transmission of packets using these routes could be higher than anticipated. In MECN, the radio model used only considers the energy spent by the transmitters and lacks the energy dissipation of the receivers. According to the work in [20], under conditions of limited transmission distances and/or high energy consumption of radio electronics, the overall energy

IoT and Topology Control

consumption of MECN can be greater than that of algorithms that use the direct transmission to the base station.

For its network life concept, SMECN uses the number of nodes remaining alive over a period of time. A simulation was performed to compare SMECN's network life efficiency over MECN. The result reveals that SMECN has more live nodes than MECN [5]. However, the measured life on the basis of this description alone is not accurate since it cannot represent the criticality of the nodes. Since SMECN prefers the direction that has the least energy routes, each node tends to send messages to the base station through the same route via its closest neighbors. The nodes along this route are the crucial ones of which, without due treatment, SMECN will overuse the resources of the critical nodes and shorten the life of the network.

Another important metric to be included in the concept of network existence is the connectivity of the network to the base station. In certain implementations, such as data monitoring, failure to send data to the base station is used to characterize the end of the network life, even if the number of live nodes is significant. In SMECN, connection to the base station is specified by the number of live nodes remaining attached to the base station. SMECN uses this term to define the network's ability to connect with the base station. Incorporating this concept in the inability to relay data to the base station will offer a more detailed definition of the existence of the network. Consider, for example, a situation in which the number of live nodes is low but the network may still provide a useful task for transmitting data to the base station. In this case, the network can be viewed as still alive and this situation can be captured with the use of both meanings.

2. **Power Mode Approach:** This section addresses the power mode algorithms by highlighting their advantages and drawbacks, as well as the network existence concept used.

GAF communication is highly determined by the density of the network and the precision of the radio node model. Connectivity and routing fidelity are ensured in dense networks by the presence of multiple communication routes. But in sparse networks, connectivity and routing fidelity are low if there is no active node in the system. GAF is a position-dependent algorithm. It depends heavily on the availability of global location information to shape virtual grids and connect nodes to grids. Although the information provided by global location information is extremely reliable, the use of global location information puts a burden on networks with minimal resources. The network life also increases proportionally with the density of the node. Extension of network life is more important in a crowded network due to the large volume of energy savings obtained by powering a variety of redundant nodes that engage in routing. On the other hand, preserving the survival of the network could not be evident in a sparse network. The GAF guarantees no contact overhead. Each node broadcasts only one message during exploration and in active states. The message above the GAF is $O(V)$, where V represents the number of neighbors of each node.

STEM uses node replication to conserve resources on the network. In the same way as the GAF, the energy savings produced by a dense network are much greater than the energy savings achieved by a sparse network. This is due to the fact that dense networks have more redundant nodes that can be exclusively converted to sleep mode. STEM believes the nodes in the network mainly remain in the monitoring state and have infrequent data forwarding operations. This presumption means that STEM is an application-specific algorithm, so STEM is energy efficient for sensor nodes that have an occasional data transfer that is caused by an incident.

STEM includes the radios of the receiver nodes and the corresponding nodes along the contact paths prior to the transmission of data. These nodes must also wait for an acceptance from the nodes of the recipient. There is also a possibility that nodes would encounter a delay that could lead to a data latency problem. The benefit of STEM is that nodes depend on local knowledge to settle on their wake-up time. The decision to transition to sleep mode is also taken locally where nodes automatically turn off their radios after data is transferred. Significant energy savings are often rendered as nodes spend much of their time in this mode. But regular intermittent switching between sleep and active listening to incoming packets normally consumes a large amount of energy. The energy usage associated with this switching phase is not defined in STEM. In addition, time, space, and message overheads for setting up and transmitting data are not suggested. The connectivity of the STEM is defined by the average number of neighbors "M" that is given in Equation (2.6),

$$M = \frac{n \ R2\pi}{L2} \qquad (2.6)$$

where "n" is the total number of nodes in the network, "R" is the propagation range of nodes, and "L" is the length of the square area. Good networking is possible for a dense network to be deployed in a specific area.

ASCENT uses node density replication to increase the life of the network, such as GAF and STEM. This means that energy savings are important as nodes are densely deployed in the network. The advantages of ASCENT are its mobility and adaptive mechanisms, which allow the parameters to be modified to suit the specifications of the applications. But setting parameters to correctly represent the specifications of applications is not an easy process that can make ASCENT impractical. The parameters concerned are the neighbor threshold value *(NT)*, the loss threshold value *(LT)*, the sleep timer value Tt, the passive timer value *Tp* and the sleep timer value *Ts*. The value *(NT)* can be modified to improve network access. By setting *(NT)* to a much lower value, the average degree would be lower, resulting in poor network bandwidth.

The low connection will split the network when the energy of the active nodes is drained. The *(LT)* parameter determines the highest amount of data loss that the network can tolerate. If the data loss rate exceeds the *(LT)* value, nodes in the passive state will be transferred to the active state to engage in data forwarding. Setting the *(LT)* value is dependent on the network application. For example, networks that are heavily mobile tend to have high data losses, while networks that are used for environmental monitoring, such as the bushfire incident, are likely to have low data losses.

The values of T_t, T_p, and T_s describe the time the node remains in the test, passive and sleep states, respectively. Choosing the values of T_t, T_p and T_s is a trade-off between energy efficiency and decision performance. For example, setting T_p to a higher value means that nodes would have more time to gather data loss information from their neighbors to make more decisions, thus leading to the high precision of ASCENT but at the cost of high energy consumption associated with changing more node radios. The higher value of T_s results in increased energy savings but fewer nodes in a passive environment. These nodes are used to back up active nodes, to connect to them when the network bandwidth is limited. Consider the situation where the network connection is very limited due to the loss of resources in the active nodes with no nodes in the passive state. The network can experience partitioning if the active nodes are the only links that bind to other nodes. ASCENT's time, message, and space complexities are not provided.

2.3.2 Evaluations Based on the Network Lifetime Definitions

The definition of the network existence in the power mode approach is not given in STEM and ASCENT. GAF, on the other hand, describes the existence of the network.

GAF uses two metrics to describe the existence of the network. The first metric calculates the existence of the network as a proportion of the surviving nodes as a function of time. This network life-time metric is used to evaluate GAF efficiency in both low mobility (1 m/s) and high mobility (20 m/s) patterns. GAF considers a number of pause periods to reflect the movements of the node. The shorter pause time represents moving nodes, while the longer pause time represents no movement of the node. GAF reveals that nodes with shorter pause periods result in a better network life than nodes with longer pause times. This is because the moving nodes may switch into grids with other nodes that allow them to share the load.

The second metric used to describe the network life is the time t before the packet distribution ratio is drastically decreased. This description tests how long the network can deliver packets effectively before the ratio drops below a certain threshold value. It tests GAF's capacity to link to the base station.

GAF uses both metrics to measure the life of the network since the use of either the first metric or the second metric alone is not sufficiently precise to reflect the life of the network. Consider using a fraction of the live nodes to define the existence of the network. As GAF's network life is closely linked to the density of the network, a small fraction of the live nodes may produce traffic without impacting routing fidelity, whereas routing can be interrupted in a sparse network. On the other hand, the second metric allows one to specify an effective packet distribution ratio. However, the GAF does not address the procedure for determining the ratio and identifying the correct packet distribution ratio.

3. **Clustering Approach:** In this section, the topic of clustering algorithms illustrates the advantages and drawbacks of clustering algorithms, as well as the network existence concept used.

Power-Aware Connected Dominating Set (PACDS) uses a simple CDS labeling method. This process offers a fast and easy way to create a network backbone.

PACDS needs one round of messages to be shared for the labeling process and one more round for the pruning process. As a consequence, PACDS can be completed in a constant number of rounds. The authors [13] stated that the time complexity of the estimation of the CDS is $O(\Delta^2)$, where 'Δ' is the maximum degree of the node in the graph. The complexity of the message is given by $O(n\Delta)$, where n is the total number of vertices or nodes in the graph [13]. However, these statements are denied in [25] where the time and message complexity of PACDS can be higher. According to [25], in the process of pruning, node u can need to investigate as many of the $O(\Delta^2)$, pairs of neighbors as possible.

Also, for each pair of neighbors, as much time as $O(\Delta)$ can be taken to figure out if such a pair of neighbors together exceeds all other neighbors of U. The time complexity of PACDS may then potentially be as high as $O(\Delta^3)$. The calculated message complexity of PACDS is $O(m)$ where m is the number of edges in the unit-disk graph, as each edge corresponds to two messages in the first stage [25]. The number of edges 'm' can be as high as $O(n^2)$. In PACDS, any time the topology changes, the backbone is reconstructed to update the changes. Frequent changes in the topology caused by highly mobile nodes can waste energy resources on the network. PACDS is also only useful for static and low-mobility networks.

Energy-Efficient Connected Dominating Set (ECDS) uses local information to obtain the desired global property. The outcome in [13] shows that ECDS nodes will live longer than PACDS. The authors [15] have shown that the message complexity of the ECDS algorithm is $O(n)$ since, in the worst case, each node sends one message during each step. The time complexity of the ECDS is also $O(n)$, which is estimated by the construction of the MIS. However, the difficulty of the ECDS message could be higher than stated in [15]. This is due to the regular exchanging of messages during the search of MIS and connector nodes. The ECDS algorithm could also not be suitable for complex networks. In the second step, the decision to select a connector node is taken on the basis of a weight metric. Therefore, MIS nodes need to compare the weight of their neighbors and consult other nodes before linking. This method can result in a high message overhead. Connectivity of the network is ensured as long as the CDS remains connected. The ECDS has proved the correctness of the ECDS in the construction of the CDS.

TMPO has many benefits. First, TMPO performs a local computation of a minimal dominating set (MDS) based on two-hop neighbor information. Second, it uses a preference parameter that takes into account the movement of the node and the energy level. As a result, a node with a higher energy level and low mobility is more likely to become a cluster head. To obtain a consistent topology, the priority parameter is used. Computationally intensive is the insecure network that needs periodic topology constructions.

Various mobility conditions have been used to assess the integrity of the TMPO-constructed topology. TMPO is also ideal for low and high mobility networks. Third, the selection of the cluster head is often rotated after a certain amount of time to disperse the load equally. TMPO's efficiency review reveals that it has greater load handling capabilities and higher maintenance topology than other heuristics. The meaning of TMPO's willingness is used to monitor network access. Clusters that are extremely mobile are likely to be isolated from their neighbors.

IoT and Topology Control

In order to prevent network partitioning, TMPO adjusts the value of willingness to a much lower value to exclude cluster heads from dominant sets. The downside of TMPO is the complexity of handling the hierarchical nature of the network. TMPO needs at most three-hop neighbor information to locate doorways and gateway nodes. Any change in the position of cluster head, host, gateway or doorway would cause nodes to propagate updates to their neighbors, resulting in a delay in updating the changes to the topology.

2.3.3 Evaluations Based on the Network Lifetime Definitions

Most of the three clustering algorithms provide the network life of their experiments, except for the TMPO algorithm.

PACDS determines the life of the network as the time before the first node in the network runs out of energy and dies. This description does not require access to the base station. If the first node to fail is a backbone node, there is a risk that the backbone column will have to be rebuilt. If the first node to fail is a node outside the backbone, the network will still run. This concept can be reasonable for a network composed of nodes that are similarly essential and the loss of a single node is not permissible.

ECDS describes the network life as the number of intervals that the network will survive until it can no longer create a CDS. In other words, the network crashes so it can no longer construct a backbone. This network life concept will characterize the efficient transmission of messages to the base station as long as the backbone remains. However, in harsh environments where nodes sometimes crash, this concept may overestimate the existence of the network. This is because ECDS cannot be fast enough to react to complex changes in the environment, resulting in a longer lifespan. Importantly, the repeated re-calculation of the backbone will absorb a large amount of energy that adds to the energy drain of the backbone nodes.

4. **Hybrid Approach:** In this section, discuss the hybrid algorithms, highlighting their advantages and disadvantages as well as network lifetime concepts.

SPAN integrates the clustering approach with the power mode approach to allow idle non-CDS nodes to switch to sleep mode, thus saving energy usage and simplifying switching mode operation. Sleeping nodes are still capable of collecting packets because SPAN is running at the top of the 802.11 ad hoc power-saving mode. As a consequence, SPAN minimizes packet delays and packet retransmissions. However, the power-saving mode feature can restrict the ability of SPAN to save energy if nodes constantly turn from sleep mode to active mode to listen to traffic advertisements.

The complexity of the SPAN message is $O(n)$ since each node exchanges one message during the notification of the coordinator or the removal of the coordinator [16]. In the worst case, the time complexity of SPAN is $O(n)$ since SPAN has to consider 'n' the total number of nodes in the network to create the CDS backbone [16]. SPAN needs to collect HELLO information from transmitted updates; thus, it relies on the routing protocol. SPAN is realistic because it does not require a location information system to determine the direction of the nodes.

In SPAN, location information is provided by the GOD module through the exchange of HELLO messages. However, the position given by the GOD module is less precise than the location provided by the location information system. In SPAN, each node must maintain a limit of three-hop neighborhood information for the notification and withdrawal of coordinators. Maintaining and upgrading three-hop information can result in a large overhead alert. The connectivity of SPAN is poor as it is controlled by the rules used during the announcement and removal of the coordinator.

CLUSTERPOW is not closely related to a particular routing protocol and can thus be used for any routing protocol. Since CLUSTERPOW is an extension of COMPOW, it can be used in a homogenous network by setting the common power to a minimum value. CLUSTERPOW does not have a leader or gateway and the clusters are created automatically when the power level is selected. This feature simplifies the cluster forming process as nodes do not need to select cluster heads or gateways. In this way, the energy resources required to select cluster head nodes and create clusters can be saved.

The architecture of CLUSTERPOW is similar to COMPOW, so its message overhead is determined by the number of power levels used in the network. In the worst-case scenario, the overhead of CLUSTERPOW is $O(Pn)$, where 'P' is the number of power levels used by nodes, while 'n' is the total number of nodes in the network. CLUSTERPOW depends on global information when deciding on the minimum power level needed for routing because each node must consult the master routing table.

The master routing table consists of input from different routing tables. The overhead message for building and maintaining multiple power routing levels is significantly high. The advantage of CLUSTERPOW is that its design has been tested on CISCO wireless cards and its accuracy has been verified. Even though the system experienced a technical problem when changing the transmit power level, they managed to test it on their laptops. The practical implementation of CLUSTERPOW is therefore proven.

The advantages of LEACH are energy efficiency and the extension of network life. LEACH is a localized algorithm that allows each node to gather information from its neighbors to form clusters. Each node sends one message during cluster setup, so the message and time complexity of LEACH is low. The message and time complexity of LEACH is $O(n)$, where "n" is the total number of nodes on the network [16]. Unlike other clustering or hybrid approaches, LEACH uses data aggregation to compress the size of the messages before they are sent to the base station. This reduces the energy involved in transmitting large amounts of data over long distances.

There are, however, several drawbacks to LEACH. First, LEACH performs a lot of tasks. The operation of LEACH is therefore quite complicated. Each cluster head node is assigned to TDMA scheduling and data forwarding tasks related to data aggregation. These demanding tasks can drain the energy of the cluster head nodes and shorten the life of the nodes. Second, cluster head nodes use the long-distance transmission to send data directly to the base station.

In other words, LEACH has a scalability problem in which it does not leverage multihop connectivity between two clusters that can lead to energy savings and network scalability. The scalability problem of LEACH was discussed in HEED [26], in which HEED suggested inter-cluster routing between cluster head nodes to enable multihop communication with the base station. Third, LEACH believes that all nodes have data to transmit, whereas in reality this presumption may not be valid, and thus energy is wasted. Fourth, LEACH randomly selects cluster head nodes according to parameters that consider the number of times the nodes become cluster heads and the fixed percentage of cluster heads. These requirements do not account for the remaining energy capacity of each node and it is conceivable that a node with a lower energy level is chosen as a cluster node. In LEACH, cluster head nodes are burdened with different tasks, so naming a lower energy node to become a cluster head will shorten the life of the network as it has insufficient energy to execute the tasks. Even if the use of TDMA-based scheduling will prevent multiple retransmissions, it is not easy to synchronize nodes.

2.3.4 Evaluations based on the Network Lifetime Definitions

In this hybrid approach, the network lifetime is defined in PACDS and LEACH but not in the CLUSTERPOW.

The network life in SPAN is defined by the fraction of the CDS nodes that remained alive as a function of time. However, an appropriate measure describing the number of nodes that must remain alive to sustain the operation of the network is not given. This description could be fair if the fraction of the live CDS nodes is capable of supplying connectivity to the base station. In a specific application, such as routing, the dominant set built-in SPAN must remain connected. Consequently, the fraction of the CDS nodes that remained alive must be connected or the routing is interrupted.

In LEACH, the number of sensors still alive is used to calculate the life of the network. LEACH also calculates the number of rounds from the first node to the last dying node. A certain energy threshold is allocated to each node. The system life of LEACH is seen to be higher than other algorithms, regardless of the energy thresholds allocated to the nodes. Although LEACH may prolong the existence of the network, it is not known whether or not the remaining nodes in the network will form a backbone. If the backbone cannot be built, the data cannot be passed to the base station. However, LEACH reveals that nodes die at random, suggesting a rational distribution of the load in the network. This function is desirable for monitoring applications in which the network will cover any region of interest since each region has at least one node to control the location.

Table 2.2 shows the comparative analysis of four types of topology control algorithms with their advantages and drawbacks for different approaches. In this scheme, the topology control methods are broadly categorized into four groups, based on parameters such as power adjustment, power mode, clustering, and hybrid.

TABLE 2.2
Comparison of different topology control algorithms

Category	Algorithm	Advantage (s)	Disadvantage (s)
Power Adjustment	Minimum Energy Communication Network (MECN)	1. Maintains energy network with low power 2. Fault-tolerant 3. Optimal spanning	1. Fault-tolerant depends upon the specific application 2. Low scalability 3. Link repair required upon topology changes
	Small Minimum Energy Communication Network (SMECN)	1. Less Energy than MECN 2. Links maintenance cost is less	1. Maximum power usage 2. No. of broadcast messages is large 3. Link repair required upon topology changes
Power Mode	Common Power Level (COMPOW)	1. Practical-based topology control. 2. Built on a wireless testbed	1. High message overhead for computing multiple power levels
	Geographical Adaptive Fidelity (GAF)	1. Optimize the performance of WSN 2. Highly Scalable 3. Maximize the network lifetime 4. Limited energy conservation	1. High overhead 2. Doesn't take care of QoS during data transmission. 3. Limited mobility 4. Limited power management
	Sparse Topology and Energy Management (STEM)	1. Energy efficient for event-triggered Applications 2. State switching	1. Trade-off energy savings with setup latency
	Adaptive Self-Configuring Sensor Network (ASCENT)	1. Self-reconfigurable and adaptive to react to applications' dynamic events	1. Possibly fast energy depletion among active nodes due to uneven load distribution
Clustering	Power-Aware Connecting Dominating Sets (PACDS)	1. Simple and quick to calculate the connected dominating set	1. Not suitable for high mobility
	Energy Efficient Distributed Connecting Dominating Sets (ECDS)	1. Node's energy residual considered in the construction of the connected dominating set	1. High message overhead
	TMPO	1. Stable topology and load balancing features. Appropriate for high mobility networks	1. High message overhead and computationally intensive

TABLE 2.2 *(Continued)*
Comparison of different topology control algorithms

Category	Algorithm	Advantage (s)	Disadvantage (s)
Hybrid	SPAN	1. Location service-free and exploits the advantage of power-saving 802.11 for routing 2. Routing Backbone	1. Nodes have to periodically wake up and listen for traffic advertisements
	Cluster Power (CLUSTERPOW)	1. Easy maintenance of clusters and possible implementation on a wireless card	1. Significant message overhead for computing multiple power levels
	Low Energy Adaptive Clustering Hierarchy (LEACH)	1. Offers a variety of energy-efficient Mechanisms	1. Complicated tasks performed by Cluster heads and not scalable

2.4 IoT AND TOPOLOGY CONTROL PROTOCOLS

This section, address the different topology control protocols namely link efficiency-based topology control, improved reliable and energy-efficient topology control, cellular automata-based topology control, and heterogeneous topology control algorithm.

2.4.1 Link Efficiency-Based Topology Control

WSN integrates the physical and computational world by analyzing environmental phenomena via ubiquitous devices called sensor nodes or motes. The Adhoc wireless sensor network consists of small autonomous devices embedded in a metric space that can detect their environment, communicate through radio broadcast, and perform local computing. The main challenge for energy-restricted wireless sensor node networks is to optimize network life. Some nodes can lose their energy more quickly or become unstable during the process. Topology control is an important technique for improving the energy efficiency of the wireless sensor network. Since the energy-efficient topology control algorithm is one of the important design parameters, the focus of this research is to develop efficient topology control algorithms to improve network performance. The proposed approach proposes a new paradigm of topology control focused on the fairness, admissibility, and effectiveness of RSSI nodes. It deals with the dynamics of removing unnecessary connections within a dense network. It increases the life of the sensor nodes, preserves the network while maintaining connectivity, and ensures that the network is linked to an effective energy link.

2.4.2 Improved Reliable and Energy Efficient Topology Control

In mission-critical applications, the loss of packets is not appropriate. It is agreed that WSN packets are connected with their neighbors, that there is a risk of packet loss, and that, in this way, reliability can be achieved while improving energy efficiency. Topology control is a powerful technique to improve the energy efficiency of wireless sensor networks (WSNs). It is an advantageous but highly complex process. If it is not carried out painstakingly, it can offer an undesired outcome. The Topology Control System: Distributed Algorithm, Local Data, Need for Local Data, Connectivity, Coverage, Small Node amount, and Simplicity [3] are important during planning. In the normal model, the network model relies on the suspicion that a few nodes are either "associated" or "detached" If all nodes are connected to the network, the network is said to have maximum connectivity. This approach is called the Connectivity-based Topology Control. Construction and maintenance of topology are two steps of topology control. In the building stage, a topological property is set up.

2.4.3 Cellular Automata-Based Topology Control

The objective of the topology control problem in WSNs is to select an appropriate subset of hubs ready to screen the zone at least at the cost of energy consumption in order to develop the system's lifetime. Topology control calculations are seen because of the determination in a deterministic or randomized system of a rational subset of sensor hubs that must remain dynamic. We use the Cellular and Cyclic Self-Reproduction Method for leading re-enactments to test these calculations and to assess the effect and part of the region to be calculated in the effective use of the calculations. Distinctive square design plans are used to **pick a district** as part of self-reproduction based on execution. The piece option is made by upgrading states over a cross-section of cells. To begin with, the cella itself is considered to be a neighborhood, but when the state of the cella is maintained by the use of moving jobs, the locale is redesigned.

2.4.4 Heterogeneous Topology Control Algorithm (HTC)

The Heterogeneous Wireless Sensor Network (HWSN) is an ad hoc wireless network consisting of a large number of different types of sensor nodes with different capabilities, such as different computing power and sensing range. Unlike the homogeneous WSN, nodes are randomly deployed in the monitored field, which is equipped with extremely different energy. Heterogeneity will significantly increase the average delivery rate and network life if nodes are deployed as an efficient network [27]. HWSN, therefore, has more practical and broader applications than the homogeneous WSN. However, due to the irreplaceability of the nodes and limitations [28], one of the major challenges is how to use energy efficiently and prolong the existence of the network in such a complex heterogeneous network.

Unlike the homogeneous WSN, there are several unique features of the HWSN. Popular forms of heterogeneity in WSN are typically listed as follows:

IoT and Topology Control

1. **Computational heterogeneity:** Different nodes have different capabilities to store information or deal with changing events. Some super or advanced nodes have a more efficient processor and memory than any regular nodes. With high-performance hardware, these nodes, which have strong computing resources, can have more and more data storage and complex data processing capabilities, e.g. when a node receives a large number of data at the same time, data fusion is performed.
2. **Link heterogeneity:** Similar to computational heterogeneity, due to power electronic devices, some nodes have more channels, a higher bandwidth, and a longer contact distance than usual nodes. They will provide a more efficient and robust data transmission network.
3. **Energy heterogeneity:** In these three common forms of heterogeneity, this is the most important and central point. Computational heterogeneity and link heterogeneity often contribute to the use of more energy than nodes in the homogeneous network so that their lifespan is rapidly reduced. Energy heterogeneity can be interpreted as which nodes are virtually equipped with different energy times.

2.5 IoT AND ROUTING PROTOCOLS

The Internet has grown exponentially over the last few decades, incorporating various implementations in many sectors, including business, travel, education, entertainment, etc. Over the years, numerous computers, services, and protocols have been developed and the Internet has evolved and continues to expand rapidly. The next generation of this digital network is the IoT, where a vast number of 'Things' are supposed to be part of the Internet, bringing new possibilities and challenges. This involves sensor nodes, radio frequency identification (RFID) tags, near-field communication (NFC) devices, and other wired or wireless gadgets that communicate with each other and with the current network providing revolutionary applications and, at the same time, presenting a variety of challenges for the research community to solve.

Wireless sensor networks (WSNs) play a vital role in the advancement and growth of IoT, empowering low-end devices with minimal resources to connect to the Internet and ultimately deliver life-changing services.

In this section, we address several standard and non-standard protocols used for routing in IoT applications. It should be remembered that we have partitioned the network layer into two sublayers: the routing layer that manages the movement of packets from source to destination, and the encapsulation layer that shapes packets.

2.5.1 ROUTING PROTOCOL FOR LOW-POWER AND LOSSY NETWORKS (RPL)

Routing Protocol for Low-Power and Lossy Networks (RPL) is a remote vector protocol that can accommodate a number of data link protocols, including those described in the previous section. It creates a Destination-Oriented Directed Acyclic Graph (DODAG) that has only one path from each leaf node to the root to which all traffic from the node will be routed. At first, each node sends a DODAG

Information Object (DIO) to advertise itself as root. This message is distributed throughout the network and the entire DODAG is steadily being set up. When transmitting, the node sends a Destination Advertising Object (DAO) to its parent, the DAO spreads to the root, and the root chooses were to transfer it based on the destination. When a new node wishes to join the network, it sends a DODAG Information Solicitation (DIS) request to join the network and the root will respond with a DAO Acknowledgment (DAO-ACK) confirmation to the network. RPL nodes may be stateless, most normal, or stateful. A stateless node keeps track of its parents only. Only root has the full knowledge of the whole DODAG. Thus, in any situation, all interactions go via the root. A stateful node keeps track of its children and kin, and thus does not have to go through the root as it interacts within a subtree of the DODAG [29].

Low power and Lossy Networks (LLNs) are WSNs in which routers and nodes are highly limited in terms of computing capability, battery, and memory size, and their interconnections are unreliable at high loss rates, low data rates, and low packet transmission rates. In comparison, they have separate traffic patterns: point to point (P2P), point to multipoint (P2MP), or multipoint to point (MP2P) [30]. Thousands of nodes can eventually be participating in these networks. As IoT is evolving, routing in LLNs is one of the main problems, the shortcomings of LLNs and restrictions must be taken into consideration when planning and implementing any routing protocols.

The ROLL (Routing Over Low-Power and Lossy Networks) working group focused on routing protocol architecture and dedicated to standardizing the IPv6 routing protocol for LLNs to fulfill the predefined specifications of four IoT implementation scenarios: home automation (RFC 5826), industrial control (RFC 5673), urban environment (RFC 5548) and building automation (RFC 5867). The ROLL working group carried out a thorough review and evaluation of the current routing protocols [31]: Intermediate System to Intermediate System (IS-IS), Open Shortest Route First (OSPF), Ad Hoc on Demand Vector (AODV), and Optimized Link State Routing (OLSR), which led ROLL to discover that these protocols did not meet the specifications of the LLNs. It is clear that the standard IP routing protocols are not capable of meeting the specifications of the multipoint application in WSNs [32,33], and hence ROLL WG argued that the IoT technologies for the transition to IPv6 were developed to include the IPv6 routing architectural structure for IoT application scenarios.

2.5.1.1 What Is RPL?

The RPL is one of the infrastructure protocols [34], a distance-vector and a source routing protocol that is built over multiple connection layer mechanisms, including IEEE 802.15.4 PHY and Media Access Control (MAC) layers [33]. It targets collection networks (WSNs) containing up to thousands of routers (nodes) where the bulk of those routers have very specific and limited resources. RPL embraces three distinct traffic flow patterns [35]: point-to-point (P2P) between nodes, point-to-multipoint (P2MP) for configuration purposes, and multipoint-to-point (MP2P) for data collection. As stated in [33], the RPL rule is to organize the WSN as a Direct Acyclic Graph (DAG) rooted at the sink node and to reduce the cost (i.e. shortest paths) of reaching the sink from every node in the WSN using an objective function.

IoT and Topology Control

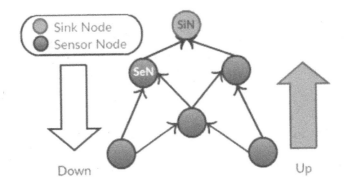

FIGURE 2.13 RPL network topology.

2.5.1.2 RPL Network Topology

RPL generates a star topology with one sink node (root) at the top and leaves at the bottom represented in Figure 2.13, leaves often attempt to enter the sink node with an Up direction (many-to-one), and down direction is devoted to traffic coming from the sink (one-to-many) [8].

2.5.1.3 RPL Messages

RPL uses four ICMPv6 control messages for creating and maintaining the routing table and the DODAG [9]:

DODAG Information Request (DIS): if a single node wishes to reach the DODAG and has not heard a DIO message for a period of time, it sends a DIS message to see if there is a DODAG that can invite it.

DODAG Information Object (DIO): used by RPL to build and manage DODAG, every node begins to send this message to its neighbors as soon as the RPL network starts. DIO provides information about the node, whether it is stored or not, and information about the DODAG setup that can support the parent's assignment process and invites unattached nodes to the DODAG.

Destination Advertisement Object (DAO): Used to disperse destination information upwards around the DODAG. The DAO could be used by a child to its parent as a request to encourage them to enter the DODAG.

DAO-ACK: a response sent by a parent to a child meaning yes or no.

2.5.1.4 Routing with RPL

In order to satisfy the specifications of the modern Low Power and Lossy Network (LLN) applications, RPL has been configured as a bi-directional routing protocol and can redirect traffic in two directions up and down, as seen in Figure 2.14.

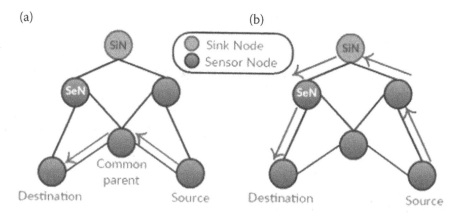

FIGURE 2.14 DODAG with storing mode (a) DODAG with non-storing mode (b) DODAG with storing mode.

1. **Upward Routing:**

Used to transfer the data obtained by the LLN nodes to the sink node, also known as the LLN boundary router (LBR), using the information on a tree-like routing topology called the Destination-Oriented Directed Acyclic Graph (DODAG), it is possible to construct the most preferred parent for each LLN sensor node by transmitting the data upwards to the parent to complete the data at the LBR.

The DODAG uses a ranking to evaluate the exact location of each node, states that the node has always a higher rank than its parents, and its estimation relies on the objective function (OF).

Three steps are necessary for the DODAG construction phase:

1. The Administrator-designed Sink begins sending multicast messages called DODAG Information Object (DIO) containing the sender rank.
2. As the children receive this message and find that it comes from a higher-ranking node, they design the sink as their relative.
3. Eventually, the other nodes will also start sending DIO messages, if a node receives several DIO messages, it will pick the best parent based on the Objective Function (OF) metrics (Expected Transmission Count, Hop count, etc.)

This method continues until all the single nodes join the DODAG, if there is a node that does not receive a DIO message and wishes to join the DODAG, it sends a DODAG Information Solicitation (DIS) to another node that has already entered the DIO message [10].

2. **Downward Routing:**

Routes are built according to the RPL Destination Advertisement Object (DAO) which includes details such as how a node will reach a number of destinations. RPL offers two types of operation storage and non-storing, both of which can be adapted to the following performance and hardware limitations:

- **Non-storing mode:** only the sink node (LBR) that knows the full DAG topology and all parent-child relationships, which means that all data must travel through the LBR before reaching any destination seen in Figure 2.14(a), this mode prevents any local routing knowledge on LLN nodes that makes it ideal for devices with very small storage capacity[10].
- **Storing mode:** LLN nodes start sending special DAO messages to their parents to reveal the paths, the node that receives a DAO message stores the sender's address, and the next hop to meet the sender. This mode can be interesting since it can reduce the overhead of the network seen in Figure 2.14(b).

2.5.2 Cognitive RPL (CORP)

The extension of the RPL is CORPL, or cognitive RPL, which is optimized for cognitive networks and uses the DODAG topology generation but has two new modifications to the RPL. CORPL uses opportunistic forwarding to forward the packet by picking several forwarders and coordinates between nodes to pick the right next hop to forward the packet to. The DODAG is designed in the same manner as the RPL. Each node retains a forwarding collection instead of its parent only and updates its neighbor by using DIO messages. Based on up-to-date information, each node dynamically updates its neighbor priorities in order to create a forwarder set [36].

In order to overcome the above-listed problems, we are designing an opportunistic forwarding approach [37] that consists of two main steps: selecting a forwarder set, i.e. each node in the network selects several next-hop neighbors, and a synchronization scheme to ensure that only the best receiver of each packet forwards it (unique forwarder selection). It has been shown that the opportunistic forwarding strategy increases the end-to-end throughput and stability (by leveraging the intrinsic features of the wireless channel) of the network, which is an important consideration for the failure of the network.

The collection of the forwarder range is a crucial problem in opportunistic forwarding. CORPL takes advantage of the current RPL parent system that includes at least one backup parent in addition to the default parent. In CORPL, each node retains a forwarder set in such a way that the forwarding node (next hop) is opportunistically picked. The formation of the forwarder package is expanded on later. CORPL uses the cost function approach to assign complex priority to the nodes in the forwarder collection. In addition, CORPL uses a basic overhearing communication scheme to guarantee a unique set of forwarders.

CORPL takes advantage of the opportunistic forwarding strategy to help high-priority latency-critical alerts that need to arrive at the gateway before the deadline, as well as to pick routes with limited intervention from PU receivers. The security of the PU transmitter is assured by an optimum propagation time for the secondary network subject to interference restriction.

As nodes engaged in spectrum sensing cannot receive/forward packets, network efficiency is thus reduced in terms of end-to-end throughput, latency, and packet loss ratios. CORPL employs two different methods to boost the average efficiency of the network under different node spectrum sensing conditions.

2.5.3 Channel-Aware Routing Protocol (CARP)

Channel-Aware Routing Protocol (CARP) is a distributed routing protocol for underwater communication. It can be used for IoT because of its lightweight packets. It considers the connection efficiency, which is determined on the basis of the historical effective transmission of the data obtained from the neighboring sensors, to pick the forwarding nodes. Two situations exist – network configuration and data routing. In the initialization of the network, the HELLO packet is sent from the sink to all other nodes on the network. In data forwarding, the packet is routed from the sensor to the sink in hop-by-hop mode. Each of the next hops is computed individually. The biggest issue with CARP is that it does not accept the reusability of the data previously gathered. In other words, if the application needs sensor data only as it dramatically improves, the forwarding of CARP data is not useful to that particular application. Improvement of CARP was rendered in E-CARP by enabling the node of the sink to save previously obtained sensory data. When new data is requested, the E-CARP sends a Ping packet that responds to the data from the sensor nodes. As a result, E-CARP decreases contact overhead drastically [23].

New distributed channel-aware routing protocol (CARP) for subwater wireless sensor networks (UWSNs) for multihop data transmission to the sink. Although CARP still reaps the advantages of cross-layer architecture (short packets for robust channel access and relay selection), the disadvantages of other solutions in this link quality are taken into account in cross-layer relay selection. Relays are chosen if they have a history of successful transmissions to their neighbors on the way to the drain. CARP also blends connectivity quality with easy topology information (hop count) for routing across connectivity holes and shadow areas. Among viable relays, the CARP node selects a neighbor with the highest residual energy that can accept a greater number of packets. CARP is often designed to take advantage of modem power control were usable by choosing transmission power in such a way that shorter control packets suffer comparable packet error rate (PER) of longer data packets. This enables nodes to pick stable and secure links: When the relay has been chosen by the initial handshake, the power is improved such that the actual transfer of data packets has a comparable chance of success to the handshake itself.

2.6 FUTURE RESEARCH DIRECTION: CONTEXT-AWARE ROUTING IN IoT NETWORKS

2.6.1 Routing in IoT

According to NIC, "by 2025 Internet nodes can reside in everyday items such as furniture, paper documents, and much more". This is achievable with the aid of a number of IoT supporting technologies. The Wireless Sensor Network (WSN) is

IoT and Topology Control

known to be the primary enabling technology for IoT. If we step in the direction of IoT, the number of sensor nodes deployed across the globe is also rising at a rapid pace. Due to rapid technological development, day-to-day sensors are getting cheaper, smaller, and more efficient. It is also possible to deploy them on a wide scale. As a result, we already have a large number of sensors deployed, and it is expected that the figures will rise exponentially in the coming time. These sensors are detecting large quantities of data from the external world. These data are further redirected to the base station to retrieve intelligence from it. Data routing in the Internet of Things depends on different factors, such as the total number of hops between the source and the destination, the nature of the connection, the volume of data, the end-to-end delay, the heterogeneous network devices, and the convergence time of the network topology.

2.6.2 Need of Context-Awareness in IoT Routing

The IoT network consists of nodes of different supporting technologies such as WSN, RFID, MANET, VANET, etc. The actualization of the IoT principle in the physical world is only possible through the introduction of these supporting technologies. Generally, the network is known to be a graph where the vertices are devices and the edges are connections to such devices. As the IoT network is a large set of heterogeneous devices and related connections, it undergoes a variety of continuous changes. These changes may occur due to a variety of reasons, such as node mobility, node death, link quality variation, etc., which may result in disconnected network topology, data loss, and higher end-to-end delays. It is therefore essential to keep an eye on these changes and to use this knowledge to improve the routing mechanism by taking the right decisions. This provides a need for context-conscious routing in IoT.

The first step in the direction of using it effectively is to understand the need for context. However, a fragmented conception of the context is not adequate; in order to allow effective use of the context, we need to have very clear visibility into the context in which our application is to be implemented. The routing programme must be aware of the active context in which the device must control the environment constantly and individually and function autonomously according to the situation.

2.6.3 Context Needed for Routing

Context can be obtained from network nodes and attached to the nodes. Table 2.3 displays the extensive listing of the background that can be obtained from the nodes and connections, as well as its use for adapting the routing mechanism in a rapidly evolving network environment.

The main characteristics of context-aware system are:

1. When the situation varies from time to time, it is important to update the context from time to time required to make the correct decision.

TABLE 2.3
Context information

Source of context	Primary context	Secondary context	Usage
Network node	Energy (If battery operated)	1. Residual energy	Residual energy gives the amount of energy left over from the specified node. It can be used to make routing decisions. If residual energy is higher, there are more chances of using the node in the data relay.
		2. Lifetime of node	A lifetime of the node gives the amount of time that the node can provide the data relay facility before it becomes useless (dead) due to the full exhaustion of its resources.
		3. Bottleneck	Bottleneck nodes have information on the less residual energy available at that node and indicate that its nearby nodes are not used to transmit the data (if possible).
	Hop count	1. Distance from destination	Hop count shows the distance of the node in the network topology from the node of the drain. The farther the hop count is the node from the drain.
		2. Delay	It's the latency of the data transfer. Delay is directly proportional to the count of hop. i.e. higher hop counts more delay in data transmission. Though hop counts influence the time needed for data transmission, they are not the only factor. There are few other considerations, such as link quality and data queue length, that also cause data routing delays.
	Memory	--	If nodes have enough memory, routing may be achieved using the storage mode, otherwise the non-storing mode is preferred. In the Storage Mode case, the routing table and the context associated with the routing are stored in the memory of the node. The context would be stored in a consolidated context server in non-storing mode.

TABLE 2.3 *(Continued)*
Context information

Source of context	Primary context	Secondary context	Usage
Queue size	1. Load	The size of the queue gives an indication of the instantaneous importance of the node load being dealt. If the queue is constantly loaded for a period of time, it causes a load balancing routing algorithm to prevent congestion.	
	2. Delay	The size of the queue specifically affects the transmission delay. Higher queue size suggests higher processing times at the node.	
Network link	Link quality	1. Expected transmission count	It represents the amount of estimated packet transmissions required to receive the packet at the destination without error. The lower the importance of the data, the higher are the chances of data distribution. If its value is infinite, the relation is assumed to be non-functional.
		2. Trust level (Packet delivery rate)	Trust level is determined using the ratio of the total packets received at the destination node to the total packets sent to the destination node.
		3. Link quality indicator (based on Received signal strength indicator (RSSI) and Signal-to-noise ratio (SNR))	It is calculated on the frequency of the signal received at the destination. Boost the signal quality by increasing the transmission success rate.

2. It takes some time to collect the context such that ample time must be provided for context-conscious processes to operate effectively.
3. When the circumstances in the network shift dynamically, it is important to keep the contextual information up to date.
4. Old context data could not be useful in taking valid routing decisions that make it from time to time for the context collecting process to be carried out.

2.7 SUMMARY

This chapter introduced the topic of sensor networks and topologies by describing the various ways in which they may be configured. Also discussed different topology control methods such as power adjustment, power control, clustering, and hybrid and their comparative analysis. Power adjustment is a strategy that decreases the energy consumption of the WSN by manipulating the transmission power of the nodes. On the other hand, power mode saves resources by shutting off idle node radios and putting the nodes in sleep mode. The third group, known as clustering approaches, conserves resources by objectively choosing a collection of neighbor nodes to create the energy-efficient backbone of the network. Finally, hybrid approaches further boost energy savings by combining a clustering approach with either power mode or power adjustment method.

This section addresses the different topology control protocols namely link efficiency-based topology control, improved reliable and energy-efficient topology control, cellular automata-based topology control, and heterogeneous topology control algorithm.

Three routing protocols in IoT were discussed in this section. RPL is the most commonly used one. It is a distance-vector protocol designed by IETF in 2012. CORPL is a non-standard extension of RPL that is designed for cognitive networks and utilizes opportunistic forwarding to forward packets at each hop. On the other hand, CARP is the only distributed hop-based routing protocol that is designed for IoT sensor network applications. CARP is used for underwater communication mostly. Since it is not standardized and just proposed in the literature, it is not yet used in other IoT applications. Finally, this chapter addressed context-aware routing in IoT networks as a future research direction.

Exercises

1. What is a low power/lossy network? How does that relate to IoT?
2. What is RPL and how does it work?
3. What are some applicable examples/real-life deployments of context-aware routing in IoT?
4. Compare different topology control methods.

REFERENCES

[1] Gupta, P. and Kumar, P. R. 2000. The capacity of wireless networks. *IEEE Transaction on Information Theory*, 46(2):388–404.

[2] Santi, P. 2005. *Topology Control in Wireless Ad Hoc and Sensor Networks*. Chapter 3, Topology Control. John Wiley and Sons, ISBN: 978-0-470-09454-9.

[3] Feeney, L. M. and Nilsson, M. 2001. Investigating the energy consumption of a wireless network interface in an ad hoc networking environment. *IEEE INFOCOM*, 3:1548–1557. Citeseer.

[4] Rodoplu, V. and Meng, T. H. 1998. Minimum energy mobile wireless networks. *ICC '98. 1998 IEEE International Conference on Communications. Conference Record. Affiliated with SUPERCOMM'98 (Cat. No.98CH36220)*, Atlanta, GA, USA, Vol.3, 1633–1639. doi: 10.1109/ICC.1998.683107.

[5] Li, L. and Halpern, J. Y. 2001. Halpern. Minimum-energy mobile wireless networks revisited. ICC 2001. *IEEE International Conference on Communications. Conference Record (Cat. No.01CH37240)*, Helsinki, Finland. Vol. 1, 278–283. doi: 10.1109/ICC.2001.936317.
[6] Li, L. and Halpern, J. Y. 2004. A minimum-energy path-preserving topology control algorithm. *IEEE Transactions on Wireless Communications*, 3(3):910–921.
[7] Narayanaswamy, S., Kawadia, V., Sreenivas, R. S. and Kumar, P. R. 2002. Power control in ad-hoc networks: Theory, architecture, algorithm and implementation of the COMPOW protocol. *In European Wireless Conference*, Citeseer.
[8] Jones, C. E., Sivalingam, K. M., Agrawal, P. and Chen, J. C. 2001. A survey of energy efficient network protocols for wireless networks. *Wireless Networks*, 7(4):343–358.
[9] Ya, X., Heidemann, J. and Estrin, D. 2001. Geography-informed energy conservation for ad hoc routing. *MobiCom 01: Proceedings of the 7th Annual International Conference on Mobile Computing and Networking*, (July 2001), 70–84.
[10] Schurgers, C., Tsiatsis, V. and Srivastava, M. B. 2002. STEM- topology management for energy-efficient sensor networks. *Proceedings, IEEE Aerospace Conference*, Big Sky, MT, USA, Vol. 3: pages - 3.
[11] Cerpa, A. and Estrin, D. 2004. ASCENT: Adaptive self-configuring sensor networks topologies. *IEEE Transactions on Mobile Computing*, 3(3):272–285. doi: 10.1109/ TMC.2004.16.
[12] Du, D. Z. and Pardalos, P. 2004. *Handbook of Combinatorial Optimization*, page 332. Kluwer Academic Publishers.
[13] Wu, J., Dai, F., Gao, M. and Stojmenovic, I. 2002. On calculating power-aware connected dominating sets for efficient routing in ad hoc wireless networks. *Journal of Communications and Networks*, 4(1):59–70. doi: 10.1109/JCN.2002.6596934.
[14] Wu, J. and Li, H. 1999. On calculating connected dominating set for efficient routing in ad hoc wireless networks. *Proceedings in 3rd International Workshop on Discrete Algorithms and Methods for Mobile Computing and Communications*, page 14. ACM.
[15] Yuanyuan, Z., Jia, X. and Yanxiang, H. 2006. Energy efficient distributed connected dominating sets construction in wireless sensor networks. *IWCMC 06: Proceedings of the 2006 International Conference on Wireless Communications and Mobile Computing*, 797–802, ACM. doi: 10.1145/1143549.1143709.
[16] Labrador, M. A. and Wightman, P. M. 2009. *Topology Control in Wireless Sensor Networks with a companion simulation tool for teaching and research*. Springer.
[17] Bao, L. and Garcia-Luna-Aceves, J. J. 2003. Topology management in ad hoc networks. *MobiHoc '03: Proceedings of the 4th ACM International Symposium in Mobile Ad Hoc Networking & computing*, 129–140. ACM New York, NY, USA. doi: 10.1145/778415.778432.
[18] Chen, B., Jamieson, K., Balakrishnan, H. and Morris, R. 2002. Span: An energy-efficient coordination algorithm for topology maintenance in ad hoc wireless networks. *Wireless Networks*, 8(5):481–494.
[19] Kawadia, V. and Kumar, P. R. 2003. Power control and clustering in ad hoc networks. *IEEE INFOCOM 2003. Twenty-second Annual Joint Conference of the IEEE Computer and Communications Societies (IEEE Cat. No.03CH37428)*, San Francisco, CA, USA, 459–469, Vol.1. doi: 10.1109/INFCOM.2003.1208697.
[20] Heinzelman, W. R., Chandrakasan, A. and Balakrishnan, H. 2000. Energy-efficient communication protocol for wireless microsensor networks. *Proceedings of the 33rd*

Annual Hawaii International Conference on System Sciences, Maui, HI, USA, vol.2, 10, doi: 10.1109/HICSS.2000.926982.

[21] Santi, P. 2005. *Topology Control in Wireless Ad Hoc and Sensor Networks, Chapter 9 Distributed Topology Control: Design Guidelines*. John Wiley and Sons.

[22] Shen, Z., Chang, Y., Cui, C. and Zhang, X. 2007. A topology control algorithm for preserving minimum-energy paths in wireless ad hoc networks. *Frontiers of Electrical and Electronic Engineering in China*, 2(1):63–67.

[23] Srivastava, G., Boustead, P. A. and Chicharo, J. F. 2003. A Comparison of Topology Control Algorithms for Ad-hoc Networks. *In* L. Campbell (Eds.), *Australian Telecommunication Networks and Applications Conference*, Melbourne: ATNAC, 1–5.

[24] Santi, P.. 2005. Topology control in wireless ad hoc and sensor networks. *ACM Computing Surveys (CSUR)*, 37(2):164–194.

[25] Wan, P. J., Alzoubi, K. M. and Frieder, O. 2004. Distributed construction of connected dominating set in wireless ad hoc networks. *Mobile Networks and Applications*, 9(2):141–149.

[26] Younis, O. and Fahmy, S. 2004. HEED: a hybrid, energy-efficient, distributed clustering approach for ad hoc sensor networks. *IEEE Transactions on Mobile Computing*, 3(4): 366–379, doi: 10.1109/TMC.2004.41.

[27] Bagci, H., Korpeoglu, I., Yazici, A. 2015. A distributed fault-tolerant topology control algorithm for heterogeneous wireless sensor networks. *IEEE Transactions on Parallel and Distributed Systems*, 26(4): 914–923.

[28] Liu, Y., Yang, C. L., Tang, W. K. S., Li, C. G. 2014. Optimal topological design for distributed estimation over sensor networks. *Information Sciences*, 254(1):83–97

[29] Winter, T., Thubert, P., Brandt, A., Hui, J., Kelsey, R., Levis, P., Pister, K., Struik, R., Vasseur, J. P. and Alexander, R. 2012. RPL: IPv6 Routing Protocol for Low-Power and Lossy Networks, *IETF RFC 6550* (March 2012), http://www.ietf.org/rfc/rfc6550.txt

[30] Vučinić, M., Tourancheau, B. and Duda, A. 2013. Performance comparison of the RPL and LOADng routing protocols in a Home Automation scenario. *2013 IEEE Wireless Communications and Networking Conference (WCNC)*, Shanghai, China, 1974–1979, doi: 10.1109/WCNC.2013.6554867.

[31] . Levis, P., Tavakoli, A. and Dawson-Haggerty, S. 2009. Overview of existing routing protocols for low power and lossy networks. Internet Engineering Task Force, Internet-Draft draftietf-roll-protocols-survey-07. https://tools.ietf.org/html/draft-ietf-roll-protocols-survey-07

[32] Tripathi, J., de Oliveira, J. C. and Vasseur, J. P. 2010. A performance evaluation study of RPL: Routing Protocol for Low power and Lossy Networks. *2010 44th Annual Conference on Information Sciences and Systems (CISS)*, Princeton, NJ, USA, 1–6, doi: 10.1109/CISS.2010.5464820.

[33] Winter, T., Thubert, P., Brandt, A., Hui, J., Kelsey, R., Levis, P., Pister, K., Struik, R., Vasseur, J. P. and Alexander, R. 2012. RPL: IPv6 Routing Protocol for Low power and Lossy Networks. IETF. RFC 6550

[34] Al-Fuqaha, A., Guizani, M., Mohammadi, M., Aledhari, M. and Ayyash, M. 2015. Internet of Things: A Survey on Enabling Technologies, Protocols, and Applications. *IEEE Communications Surveys & Tutorials*, 17(4):2347–2376. doi: 10.1109/COMST.2015.2444095.

[35] Sheng, Z., Yang, S., Yu, Y., Vasilakos, A. V., Mccann, J. A. and Leung, K. K. 2013. A survey on the ietf protocol suite for the internet of things: standards, challenges, and opportunities. *IEEE Wireless Communications*, 20(6):91–98. doi: 10.1109/MWC.2013.6704479.

[36] Aijaz, A. and Aghvami, A. H. 2015. Cognitive Machine-to-Machine Communications for Internet-of-Things: A Protocol Stack Perspective. *IEEE Internet of Things Journal*, 2(2):103–112. doi: 10.1109/JIOT.2015.2390775.

[37] Biswas, S. and Morris, R. 2005. ExOR: Opportunistic multi-hop routing for wireless networks. *SIGCOMM '05: Proceedings of the 2005 conference on Applications, technologies, architectures, and protocols for computer communications*, 133–144. 10.1145/1080091.1080108

3 Design Issues, Models, and Simulation Platforms

3.1 TOPOLOGY CONTROL DESIGN ISSUES

The succeeding facets need to be considered meticulously in the design stage to attain a reliable and scalable WSN model to recover topological challenges:

1. **Energy conservation:** One of the main objectives of the design is to use this restricted energy to save energy as effectively as feasible. Energy efficiency is particularly significant in WSNs although batteries are usually unfeasible to replace or recover sensors. When power-saving strategies are employed at various levels of wireless architecture, both the individual units and the network can extend their functional lives significantly.
2. **Limited bandwidth:** Nodes that are normally wireless multi-hop networks have very little bandwidth. Although industry standards like IEEE 802.11 have a theoretical bandwidth up to 54 Mb/sec, because of radiation inference due to simultaneous communication the situation is much worse in practice.
3. **Unstructured and time-varying network topology:** Network nodes can be organized in the deployment region arbitrarily; thus, the shaped graph is usually unstructured, representing communication links. The topology of the network can also differ because one or more nodes' mobility and/or failure is caused. The proper value of basic network parameters is therefore an arduous task (for example, the essential connectivity transmission range).
4. **Low-quality communications:** The consistency of communication on wireless media is usually much lower than on wired media and environmental factors are likely to have a strong impact on the quality of communication.
5. **Operating in hostile environments:** The WSNs are designed to use in harsh environments in some situations, so the sensors have to be specially configured in severe conditions, resulting in a possible event for the individual unit failure. Therefore, sensor resilience defects at various network layers should be clearly defined.
6. **Data processing:** Due to energy restraints and the anticipated inadequate quality of communication, sensed data need to be condensed or combined with nearby sensor data before it is sent to the gateway nodes.
7. **Scalability:** As previously reported, WSNs could consist of several thousand sensors, depending on the scenario. The scalability of the protocols proposed is therefore a major task.

DOI: 10.1201/9781003310549-3

3.1.1 Taxonomy of Topology Issues

Issues of topology in WSNs were studied extensively. In this portion, a consistent taxonomy is arranged and shown in Figure 3.1. Topology problems are broken down into dual classes: One is topology awareness problems and the other is topology control problems.

Topology awareness problems consist of geographical routing challenges and sensor hole challenges. For optimum routing methods, geographical routing utilizes high routing efficiency and low energy consumption geographical and topological network details. In a WSN it is possible to establish many sensor holes, such as jamming holes, sink/black holes, and wormholes, thus generating varying network topology, which can obstruct the application of the upper levels. For example, because of intense communication and the transmission of the message to the outside node fails, jamming holes may occur. Extended nodes across a sink node or malicious nodes are responsible for sink/black holes. If sensor hole problems aren't treated correctly, they can generate an expensive routing table and quickly deplete the intermediary nodes.

The topology control problem is further categorized into two types: topology of sensor coverage and sensor connectivity. The sensor coverage topology shows the topology of the coverage and is concerned about the maximization of a stable sensor field with lower power consumption. On the other hand, the topology of networking highlights network connectivity and analyses message collection and distribution within the network. To maintain an effective sensor connectivity topology, two mechanisms were used, one is power control method and the other is power management method. The first monitors the radio capacity level and then has a strong wake-sleep schedule to achieve optimized connectivity topology.

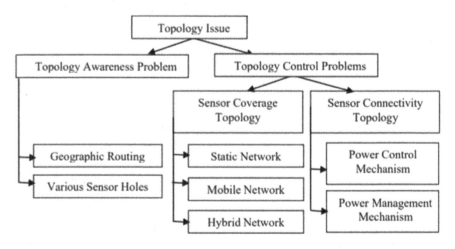

FIGURE 3.1 A taxonomy of topology issues in WSNs.

3.1.2 Topology Awareness Problem

1. **Geographic routing:** Geographic routing is based on greedy forwarding of local information on the topology of networks to route packets. The greedy perimeter stateless routing (GPSR) for MANETs is proposed by Karp et al. [1]. The protocol begins in greedy furthering mode and assumes that sensor node locale information can be achieved across systems [2,3]. By using perimeter roughing mode and the right-hand law, GPSR recovers from the local maximum location. Kranakis et al. [4] propose an algorithm for compass routing and FACE-1 which ensures destination is achieved even in greedy transfers, if the local minimum phenomenon occurs as in [4], the FACE-2 routing algorithm is proposed by Bose et al. [5]. Unlike GPSR, FACE-2 routing occurs across each node's perimeter of the Gabriel graph (GG). The FACE-1 often changes such that when the edge obstructs the line from the source to the target, the perimeter passes through the next edge.

The drawback of FACE-2 is that the perimeter nodes consume more resources. Kuhn et al. [6] implement a deterministic retrofit function as an expansion of the compass routing algorithm to return to the greedy mode of the perimeter routing mode without actually exploring the entire face boundary. In [7] an intermediate node forwarding (INF) probabilistic solution has been suggested. During this method, a negative acknowledgment (NAK) was introduced to give the source feedback on packet drops. Li et al. [8] suggest an active message transmission in a detached mobile ad-hoc wireless network via a relaying communication system. The protocol relays messages by transferring nodes to carry out a routing path on a detached network with mobile agents. The protocol projected for WSNs can escape the routing hole due to the sparse disposition or node flops, but if a routing hole is created, it will not achieve significant results. The geographical and energy aware routing (GEAR) proposed to push a packet on a region of interest was suggested by Yu et al. [9]. GEAR works well if a small portion of the overall area is to be protected by the region. As the area of interest grows, productivity falls rapidly. Table 3.1 summarizes and shows a description of the proposed geographical routing schemes.

2. **Holes Problems:** The routing hole for an area in the network, which is not the nodes, or where the accessible nodes cannot join in the definite routing of the data for numerous probable reasons, refers in most regional routing schemes. The geographic routing methods are mentioned in Table 3.1. It doesn't include the methods for detecting and locating trousers to prevent infection with sensor holes in packet delivery. The theoretical work on the determination of sensors is performed with the definition of the so-called stuck node and an algorithm known as BOUNDHOLE to identify the holes using the robust stuck nodes [10]. Li and Liu [11] are studying an underwater control application scenario in the coal mines. They recommend a SASA Topology Maintenance Protocol to speedily detect the structural variability during underground failure by adjusting the network deployment of mesh sensors and by developing a collaborative mechanism based on a standard sensor beacon strategy.

TABLE 3.1
Geographic routing schemes

Protocol	Maintained State	Capability of Topology Adaptive Fault Tolerance
GPSR [1]	Location info and the whole planar graph (RNG or GG)	Greedy transmission mode and Right-hand rule in perimeter mode to round the voids.
Compass Routing [4] FACE II [6] GOAFR+ [5]	Location info and the whole planar graph (GG) i.e. Gabriel Graph	Using face routing on planar graph to prevent routing holes.
INF [7]	Location info	Active NAKs and source-initiated repair.
Active Message Relay [8]	Location info	By moving nodes to reach disconnected neighbors.
GEAR [9]	Location info and learned an estimated cost values	The geographical routing learned and estimated cost of efficient energy and limited flooding in the region.

The so-called edge nodes describe and report the sensor hole to the sink. The first work on the topology variance with real geographical changes was the SASA protocol. Wood et al. discuss in [12] the Jamming hole. A jamming hole bypasses nodes' ability to communicate/sense within a certain area, to create a virtual hole. Wood et al. offer a JAM protocol for the sensor network detection and mapping of jammed regions. To differentiate between jamming and usual interference, the detection portion of the protocol uses heuristics based on accessible data, such as bit error rates, etc. The JAM protocol pretends that each node has locality details and a distinctive ID.

Sink/black holes and wormholes are created progressively because of the depleted strength of a sensor node and potential denial of network service attacks. The sinkhole is marked by intense resource disputes for limited bandwidth and channel access among adjacent nodes [13]. Another kind of service-attack denial is a wormhole [14]. It is formed by creating the belief that a malicious node is a neighboring node that leads to wrong routing convergence in different networks. Karlof et al. [15] investigate the stability and the conservation of topology maintenance algorithms in different protocols for routing energy toward sinkholes. They demonstrated that common routing protocols such as directed diffusion, rumor routing, and the directed diffusion multi-path variant, etc, are all vulnerable to sinking hole attacks.

For the geographical transmission of greed, the algorithms are harder to generate sinkholes since a malicious node needs to advertise to different neighbors' different desirable locations to qualify as the next hop. Several potential defenses against the sinkhole were defined by Wood et al. in [13]. Only approved nodes can share routing data in the authorization solution. Due to the high overhead of calculation and coordination, the solution is not scalable. In addition, due to the capacities and limitations of sensor systems, public-key encryption in sensor networks is not feasible.

The "Colored Based Topology Control" (CBTC) algorithm [16] is an added way of removing the sensor hole. The color of the node is denoted by the algorithm, i.e. a "color flag", in the sensor node code field. For any specific application, all sensor nodes are sensitive, e.g. humidity, temperature, or pressure. Another node can sense a node with a similar color flag in the coverage, and communication is possible between them. As we know in the sensor network when a node is dead on its route to a sink node, a coverage hole occurs. Other nodes can't pass their data to the sink node through this coverage hole. If the node of the color flag "blue" is dead when a color node leaves it with the color-dependent topology algorithm then it should operate by selecting the node of the additional color flag inside its region of coverage.

3.1.3 TOPOLOGY CONTROL PROBLEM

1. **Sensor coverage topology:** It is further divided into various groupings like static network, mobile network, and hybrid network.
    a. **Static network:** Certain methods for a static sensor network have various coverage objectives. These methods are addressed separately.
        - **Partial coverage:** Ye et al. [17] propose PEAS, which extends the operating time of WSNs device by maintaining several sensors that are required if the deployment rate of nodes is much greater than needed. PEAS consists of two algorithms: Adaptive sleeping and Probing environment. The PEAS protocol does not require pre-knowledge of node location information. The near-optimum rotating sensory coverage for the WSN surveillance system is being developed by Cao et al. [18]. The aim of their scheme is to cover the sensing area partially with each point finally felt within a limited delay. They presume that the neighboring nodes have synchronized clocks roughly and they know each other's sensing nodes.
        - **Single coverage:** The optimal geographical density control (OGDC) protocol was suggested by Zhang et al. for the single coverage requirement [19]. The protocol attempts, for cases with RC to 2Rs where RC is the communication node range and Rs the sensing node ranges, to reduce the overlap in sensing areas in all sensor nodes. OGDC is a completely localized algorithm, but it is necessary to locate the node in advance.
        - **Multiple coverage:** Wang et al. [20] proposed a coverage configuration protocol (CCP) that can deliver consistency to configure a different degree of the coverage sensor network. The CCP protocol requires knowledge about the location of the node. Huang et al. [21] proposed polynomial-time algorithms to authenticate that at least the necessary node number covers every single point in the target region. The authors propose an organization that can collect the data

of inadequately protected segments and sends additional nodes. But there is no scalability in this centralized approach [22].

b. **Mobile network:** The use of movable sensor systems is studied by Wang et al. [23]. Because of a region to monitor, the recommended self-deployment protocols initially detect the presence of coverage holes in the region to be controlled, then measure the target positions to reduce coverage holes and transfer sensors. The diagrams [24,25] of the Voronoi coverage holes are used and the protocols VEC, VOR, and Minimax for the deployment of the sensor are built in three movements. As for virtual units, Heo et al.' [26] and Howard et al. [27] are studying the sensor network. In [27] nodes only make moving decisions using their sensed information. It is cost-effective and no coordination between nodes or data is necessary. Sensors are initially randomly deployed for the DSS (distributed self-spreading) algorithm recommended in [26]. They begin to move based on neighbors' partial powers. The forces exercised by its neighbors in each node depend on the local deployment density and the distance from the node to the neighbor.

c. **Hybrid network:** The active field of research involves the coverage scenario now a few days with just a few of the capable sensors, in particular in the field of explorative robotics. Sensors that will move can assist in network maintenance and deployment, switching to suitable areas in the field to reach the necessary coverage. The joint approach for the exploration and coverage of a certain target region was suggested by Batalin et al. [28]. With the assistance of a consistently moving robot in a given target area, the coverage issue is fixed. The algorithm does not take into account the communication between the designed nodes. The robot makes all decisions by interacting directly with a nearby sensor node. Wang et al. [29] overcame the single coverage issue by shifting the available mobile sensors to heal coverage holes in a hybrid network. Table 3.2 lists a summary of various sensor coverage approaches. The majority of proposals include node location information, as can be seen from the table, as assistance and the unit disc model is commonly adopted in terms of simplifying the model of the node transmission.

2. **Sensor connectivity topology**
   a. **Power control mechanisms:** The objective of power control mechanisms is to dynamically adjust the transmission range of nodes to preserve certain properties of the communication graph while reducing the energy that node transceivers use as they are a simple energy source in WSNs. The mechanisms for controlling power will essentially achieve a good energy efficiency network. The power control study shall be carried out in homogeneous and non-homogeneous situations, which can be tested if nodes have the same transmission range. The CTR (critical transmission range) problem was theoretically and discovered

TABLE 3.2
Coverage approaches

Categorization	Approach	Features
Static network	Partial coverage	Sleeping schedule is distributed
		Guarantee finite delay bound
	Single coverage	Energy consideration is residual
		Coverage calculations are sector-based
		Uniform disk sensing model
	Multiple coverage	Configurable degree of coverage
		Supports non-unit disk model
		Disk-based differentiated degree of coverage
Mobile networks	Computational geometry	It is localized, scalable, and distributed
		Energy consideration is single coverage-based residual
	Virtual forces	Scalable, distributed, local communication not required
		Scalable, distributed. Residual energy-based
Hybrid networks	Single mobile sensor	No multi-hop communications, distributed
	Multiple mobile sensor	For single coverage requirements, Voronoi diagram is used

in a homogeneous network. COMPOW [30] is a distributed protocol that enables the minimum common range for transmitting network communication to be identified.

It shows that the adjustment of the range of transmission to this value optimizes the network bandwidth, reduces access contention to the wireless channel, and reduces the power consumption. During the interaction graph, the compromise between the transmission range and the largest component size was investigated. The experiment's findings show that the transmission range can be greatly reduced in sparse two-dimensional and three-dimensional networks when weaker communication specifications are appropriate. The largest linked components have an average size of about 0.9 n, having the critical transmission range. It means that a significant amount of energy is spent connecting relatively few nodes. Non-homogeneous networks are more difficult because nodes can have different transmission ranges. The range assignment (RA) problem is the problem of assignment of a node transmission range so as to bind the resultant communication graph firmly and come up with a minimum energy cost [31].

In the case of 2D and 3D networks, it is defined as NP-hard. However, with the created range assignment, the optimal solution can be approximated by a factor of 2. A significant RA variant was recently considered based on the hypothesis of communication graphs of symmetry. The high overhead required for handling unidirectionally linked protocols and MAC protocols that are normally intended to operate underneath the symmetric assumption i.e. symmetric range allocation (SRA) shows more practical importance. NP-hard, however, also requires SRA in 2D and 3D networks, and it still costs significant additional energy around RA. I can refine SRA in weakly symmetric range assignment (WSRA), which diminishes the requirement that communication charts contain bidirectional connections, but requires the symmetrical subgraph that results from the RA connection.

With the release of WSRA challenge, only a marginal energy cost consequence has been caused while the preferred symmetry is still present.

b. **Power management mechanisms:** Power management is concerned with the collection of nodes to be enabled and when to extend the network's lifetime to establish energy conservation topology. The protocol stack will use the information available from all layers. In the GAF approach [32], nodes use field fragmentation knowledge in fixed square grids. Regardless of node density, each grid's size is constant. Nodes in a grid turn from sleep to listening, ensuring that a particular node in each grid is left in order to forward packets with a complex routing backbone. For multi-hop wireless networks that choose coordinators adapted from the entire nodes to establish a routing backbone, a power-saving maintenance algorithm known as Span [33] is utilized to save power from the other nodes' radio receivers. A power-saving topology of the active STEM method [34] is used to leverage the time dimension rather than the node density dimension to monitor. Provide switching of nodes between the "transfer" and "monitoring" states. Data are only transmitted in the transition state. In the monitoring state, nodes hold their radio away and turn to a transmission state on an occurrence that is detected as an initiator node. The additional research on STEM and GAF integration demonstrate how power savings can further be achieved by taking advantage of both the time and node density. Table 3.3 outlines power management mechanisms and offers a detailed understanding of the characteristics of the recommended mechanisms.

3.2 NETWORK MODELS

3.2.1 HOMOGENEOUS MODEL

Suppose that n nodes are placed in R = [0,1]d for d = 1,2,3, the question arises, "What are the minimum transmitting range 'r' values that generate the linked network with a homogenous range assignment?". This minimum value 'r' is known as the critical transmitting range C_{TR} for connectivity. The critical transmission (TR) range is important to research because the transmission range cannot be dynamically changed in a variety of cases. Some expensive broadcasting transmitters may not be able to adjust the transmission range. The same range of transmission for all nodes is fair in such circumstances and the CTR is the only solution to reduce energy consumption and increase network capacity.

TABLE 3.3
Approaches of power management mechanism

Protocols	Synchronization	Mobile/Static	Distributed	Location info
GAF [46]	None	Mobile	Yes	Yes
Asynchronous Wakeup Protocols [47]	None	Static	No	No
Power Saving Protocol [48]	None	Mobile	Yes	No
SPAN [49]	None	Static	Yes	No
STEM [50]	None	Static	Yes	No
S-MAC [51]	Yes	Static	Yes	No

The importance of CTR depends on the available knowledge about the deployment of TR nodes. If the location of the nodes is determined earlier, the CTR is the longest edge of the euclidean minimum distance tree (MST). However, several WSNs are deployed in an ad hoc and placements of nodes are not known in an early stage. In the case of node placement, the minimum value of r to ensure compatibility is not known in any case, considering that nodes may be located at opposite ends of the deployment area.

The example, where nodes are deployed to the opposite ends of the deployment area, is unlikely, CTR was therefore analyzed based on the distribution of TR-nodes in R based on the certain distribution of probabilities. The objective in such a situation is to identify the minimum value of r, which provides a high possibility of connectivity. A common network power ensures two-way connections, which in turn ensures the efficiency of the MAC protocol. After reducing transmission capacity, Figure 3.2 shows the sparser topology.

3.2.2 Wireless Propagation Model

The wireless signal is transferred from source to destination through a wireless radio channel in wireless sensor communication. Signals that are forwarded by the transmitter at some power would be affected by the signal attenuation issue on a radio channel. If the received signal with a capacity level exceeds its transceiver sensitivity, communication at its destination for transmitter transmission has attenuation. Attenuation of signals is nothing but the loss of channel paths directly dependent on the distance between the source and target, other associated details and operating frequency.

3.2.3 Model of Long-Distance Path

The model of long-distance path is given by

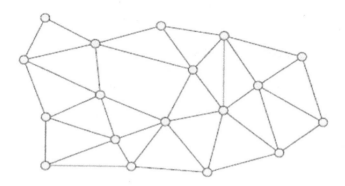

FIGURE 3.2 Sparser topology after reducing transmission power.

$$Prx(d) \alpha\ Ptx/D^\alpha$$

This says that path loss over the distance d is directly proportional to Ptx (transmitting power).

3.2.4 Hop Model

We shall address the definition and design of the hop model in this section. For every sensor node on the network, let us say p, it represents its deviation from the optimum relay location in the adjacent node position. The optimum relay location is nothing more than the direct connection between the base station and node p. This is d_{char}, the characteristic distance function. Here, node p is a coordinate system origin, u and v are represented as coordinates of node p denoting d as $d_{char} \cdot a$ and $d_{char} \cdot b$ simultaneously, where a, b ∈ A.

The total distance from node m to n is then computed as $d(m, n) = d_{char} \cdot c$, where $c^2 = a^2 + b^2$. The addressing of the network's multi-hop link depends only on local data. The optimized number of hopes needed can be calculated as,

$$K = D/\bar{a} \cdot d_{char} = K_{opt}/\bar{a}, \text{ where}/\bar{a} \text{ is the average value of all}a.$$

Different methods produce different results for topology control. Let V be graph G with nodes. If node m can enter node n directly from node m to sensor node n. Topology results should be calculated and compared with other topology control methods. A list of efficiency metrics such as energy efficiency, accessibility, robustness, performance, mobility, etc. is used for comparative study of topology control methods.

3.2.5 Energy Model

It is necessary to incorporate a model to drain energy from the nodes when the node takes some action to evaluate the lifecycle of a certain topology. To model the energy consumption of nodes, the energy model is used. It is based on the following equation:

$$E_{Tx} = E_{elec} + E_{amp} * R_{comm}^2 * \pi$$
$$E_{Rx} = E_{elec}$$

where E_{Tx} is the energy spent to transmit 1 bit, E_{Rx} is the energy to receive 1 bit, E_{elec} is the energy used by the electronic components of the radio, and E_{amp} is the energy used by the radio amplifier. The second term is proportional to the square of the radio signal propagation range. It is still widely used in wireless sensor networks literature, despite the simplicity of this energy model.

$$\text{Initial energy source} = 1 \text{ Joule}$$
$$\text{Eelec} = 50 \text{ nJ/bit}$$
$$\text{Eamp} = 10 \text{ pJ/bit/m}^2$$

Energy consumption is negligible.

3.3 SIMULATION PLATFORMS

3.3.1 OMNeT++

Name of Simulator: OMNeT++
Platform: Windows, Linux, Unix, MAC.
Developer/Licencing: Free for academic AndrásVarga (OMNeT++ Community)
Description:
OMNeT++ is a comprehensible, modular C++ simulation and modular discrete network simulation platform for object-oriented purposes primarily to construct network simulators. It has a generic architecture so that it can be used for several domains:

- Wired network modeling and wireless network modeling
- protocol modeling
- Queueing Network modeling
- Multi-processor modeling and other distribution hardware structures
- Hardware architecture validation
- evaluate complicated software device performance aspects
- Generally, any device that fits the distinct event method can be modeled and simulated conveniently into entities that communicate via message exchange.

OMNeT++ itself is not a realistic simulator but offers an infrastructure and the tools needed for simulation writing. A component design for simulation models is one of the essential ingredients of this infrastructure. Reusable parts called modules are used to assemble models. Well-written modules, like LEGO blocks, can be reused and mixed in different ways.

The modules can be linked via gates (other systems will call them ports) with each other and can be combined to form composite modules. The depth of the nesting module is not limited. Modules communicate via message passing, where messages can hold arbitrary data structures. Modules may transfer messages via predefined paths or directly to their destination through gates and links; wireless simulations are useful for the latter, for example, modules may have the parameters to customize the behavior of the module or to parameterize the topology of the model. Simple modules are referred to as modules at the lowest module hierarchy level and encapsulate model behavior. The C++ programming of simple modules allows the use of the simulation library.

Different user interfaces allow for the running of OMNeT++ simulations. Graphic, animated users are very useful for presentation and debugging, and batch execution is best done using command-line user interfaces.

The simulator is extremely portable, as are user interfaces and software. They're checked on most popular operating systems (Linux, Mac OS/X, Windows) and compiled from or on most Unix-like operating systems, after trivial modification.

OMNeT++ supports simulation that is distributed in parallel. OMNeT++ can employ many communication mechanisms between distributed parallel simulation partitions, such as MPI or labeled pipes. The parallel simulation algorithm can easily be extended, or new ones can be plugged in. Models should not be controlled in parallel by special devices, it is just a matter of configuration. Even OMNeT++ can be used to present parallel simulation algorithms in the classroom since simulations can even run under the GUI in parallel.

OMNEST is the OMNeT++ version that is commercially supported. OMNeT++ is only available for academic and non-profit use; OMNEST licenses from Simulcraft Inc are required for commercial purposes.

3.3.2 NS2

Name of Simulator: NS2
Platform: Windows, Red Hat Linux, Ubuntu OS
Developer/Licencing: Open Source and Free

Description:
Network Simulators (version 2), also known as NS2, are just an event-based method that has been used to research the communications network's dynamic existence. NS2 can be used for the simulation of functions and protocols for both wired and WiFi networks (e.g. routing algorithms, TCP, UDPs). Generally defined and simulated by NS2 are the methods for users of these network protocols.

NS2 has become increasingly common in the networking research community since it was established in 1989 because it is flexible and modular in design. Several revolutions, thanks to the substantial contributions by the players on the field, have marked the growing evolution of the tool. These include the University of CA and Cornell University, the basis upon which NS is invented. The REAL network simulator has been developed. The Defense Advanced Research Projects Agency (DARPA) supports the creation of NS using the Virtual Inter Network Test Bed

Simulation Platforms

(VINT) project since 1995. The Ride is currently being further established by the National Science Foundation (NSF). Last but not least, the community research team works constantly to keep NS2 solid and scalable.

NS2 gives the username of a Tcl simulation scripting file an executable command "ns", taking an input statement. In certain cases, it is used to draw a graph and/or generate animation by creating a simulation trace file.

NS2 consists of two languages: CCC and Command Language Object-oriented tool (OTcl). The CCC specifies the internal mechanism of the simulation (i.e. backend), whereas the OTcl configures the simulation by assembling and configuring the artifacts and planning discreet events (i.e. a frontend). The OTcl and CCC are linked by TclCL. The variables in the domains of OTcl are often called handles when mapped to a CCC object. A handle is a string in the OTcl domain (e.g. "_o10") and it has no features. Rather, in a mapped CCC object the functionality (e.g. receiving a packet) is specified (e.g. of class connector).

Features of NS2
1. It is a discreet network analysis event simulator.
2. It supports the simulation of various protocols such as TCP, FTP, UDP, https, and DSR.
3. It simulates wired and wireless networks.
4. It is primarily Unix-based.
5. Uses TCL as its scripting language.
6. Otcl: object-oriented support
7. Tclcl: C++ and OTcl linkage
8. Discrete event scheduler

3.4 SIMULATION USING MATLAB FOR IoT DOMAIN

It is the process-based language of the fourth generation. MATLAB enables user interfaces, the manipulation of matrixes, the plotting of graphs, and interfaces with other written programs such as C, C++, and FORTRAN. The world MATLAB is coined from two words matrix Laboratory. It's the technical computer language. It's a marketable commodity. In the late 1970, Cleve Molar established MATLAB. He was the chairman of New Mexico's computer science department. MATLAB was rewritten in C and Math Works was created in 1984. These reprocessed libraries were referred to as JACKPAC. In 2000, MATLAB was rewritten to manipulate the matrix with the newer collection of libraries. It was originally designed to solve problems in the form of linear algebra. You can use it easily. It can accommodate a large variety of sizes. The algorithm is sensitive to numbers. A wide range of digital/image library functions and independent platforms. Menus and buttons are easy to build GUI. Object-focused programming simplifies life.

3.4.1 THE MATLAB SYSTEM

The following points address the five main components of the MATLAB subsystems:

- Environment development: It is a tool and functionality platform that helps us use MATLAB features and files. GUI is provided for the most part in MATLAB tools. It includes the command window, the MATLAB desktop, history of command, debugger, editor, workspace, etc.
- Mathematical Function Library: This library is, as the name implies, a list of many mathematical functions and algorithms, including primary operations (cosine, volume, sum, complex arithmetic, etc.), modern operations (fast Fourier transforms, matrix inverse, Bessel functions, matrix eigenvalues, etc.).
- LAB Language: LAB language is a matrix high-level language with function, flow control, input-output, data structures, and object-orientated coding features. LAB language includes a function. This allows both small and large activities, such as programming, to support rapid and complex application programs.
- Graphics: To display matrixes and vectors as graphs with correct labeling and annotation, MATLAB is extensively supported by graphs. It consists of high-level operations for data, animation, image processing, and presentation graphics for 2D and 3D visualization. This takes into account low-level operations to fully optimize the appearance of graphics and the construction of whole GUIs.
- The MATLAB API: The MATLAB API allows us to type library programs FORTRAN and C, interacting with MATLAB software. This involves the complex link to call the routines from MATLAB. MATLAB is named for reading and writing MAT files as a computational engine.

3.4.2 MATLAB for IoT Domain

MATLAB offers thousands of functions, including predictive maintenance, signal and image management, feedback and supervisory control, optimizing, and machine learning, for the development of IoT applications.

Develop algorithms much quicker using, customizing, or creating your functionality with MATLAB than with conventional programming languages. It can work in many different IoT situations, like streaming or big data.

3.5 FUTURE RESEARCH DIRECTION: HETEROGENEITY OF NETWORK TECHNOLOGIES

The heterogeneous network is an emerging area of research with great potential to change both our considerate of key concepts of computer science and our forthcoming lives. Increasing numbers of fields, including smart home, smart city, smart transport, environment monitoring, protection, and advanced manufacturing, are being utilized by the heterogeneous IoT network (HetIoT). Therefore, HetIoT is packed with our lifespan and offers a range of suitable resources for our forthcoming in connection with strong applications. IoT's network architectures are inherently heterogeneous, with the network of wireless sensors, wireless trust, and networking facilities, wireless mesh, mobile networking, and mobile communications. Smart devices use effective

Simulation Platforms

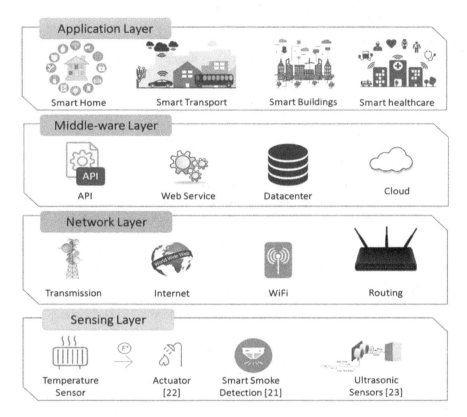

FIGURE 3.3 Future heterogeneous network IoT architecture.

communication methods in each network unit to combine digital information and physical objects that offer innovative and exciting applications and services for users.

IoT is a dynamic and multi-heterogeneous mechanism [35]. As shown in Figure 3.3, the HetIoT features four-layer architectures, including applications layer, cloud computing layer, networking layer, and sensing layer applications. The independent functioning and scalability of each layer of the four-layer architecture. Cloud servers store the sensing data composed of several sensors through efficient heterogeneous networks. There are several different network configurations in the various networking units. HetIoT has been used in everyday life and industry due to advances in the computer hardware design of sensors and the optimization of network topology.

3.5.1 Sensing Layer

Inside the HetIoT architecture sensor layer, different sensors offer expressive sensing data for the cloud server, which are sensors, color sensors, flicker sensors, motion sensors, cameras, and so on. In monitoring areas, a wide range of sensors are used and topology is made in the form of self-organizing and multi-hop sensing information. The sensor nodes sink nodes and control nodes in a typical network sensor system. Multiple hops transmit detection data from sensor nodes over sink nodes. Users run the network sensor and free the control nodes for monitoring tasks.

Some nodes are more likely to fail because of their effect on the atmosphere and energy depletion, which also creates changes in network topology. To ensure networks coverage and synchronization, to obtain efficient network topology for data transfer, redundant wireless communication lines must be eliminated by means of power control and backbone node selection.

3.5.2 Network Layer

The networking layer in the heterogeneous IoT network is used to create an effective topology to transfer data from the source node to the destination node. Networking models are also provided with a high level of data transmission for nodes, they are star network, tree network, scale-free network, and hybrid network. In addition, these network representations can forward the data through sink nodes, supernodes, or extra relay units to the cloud server. Effective topology building mechanisms may also allow network models to manage the nodes.

According to different types of HetIoT routing protocols, networking models have some drawbacks such as consumption of power, data performance, and malicious attacks. The routing protocols that self-organize boost network model robustness to withstand a part of the failure of nodes. To transfer the data to the cloud server, HetIoT requires high data transmission power. Energy-saving protocols are implemented without power in such harsh environments to prolong the life cycle of HetIoT.

3.5.3 Cloud Computing

Cloud computing and new technologies HetIoT can manage big data efficiently and reliably. HetIoT will obtain and process data from further layers in the cloud computing layer in the future [36]. Cloud servers have high analysis ability, and can also make decisions based on analytical findings in accumulation to storing data. Cloud servers can react rapidly based on emergency event awareness techniques in some HetIoT emergency applications. Rising heterogeneity of data needs intelligent decision-making utilizing efficient cloud computing. Cloud computing, because of its strong data analytics, has greater capacity in IoT compared to middleware. To provide a high quality of service for various applications, middleware can protect the discrepancies between different operating systems and dissimilar network protocols. However, a proprietary protocol is used by most common services, and interoperability is hard to achieve. In addition, due to the unsuited protocols in subsystems, middleware services have a time delay and overhead memory. The cloud server will communicate with heterogeneous systems as an abstract layer.

3.5.4 Application Layer

There will be many implementations in the application layer of forthcoming HetIoT, including MCN, vehicle networks, Wi-Fi, and WSN. Users use intelligent mobile systems, such as WeChat, Skype, Line, etc. to communicate with one another through MCN everywhere. Vehicle networks are used for the monitoring of emergency traffic incidents in intelligent transport systems. People and vehicles are linked to additional smart mobile devices and vehicle networks can forecast traffic using the cloud computing layer on the basis of real-time data traffic. Wi-Fi can provision different network connectivity protocols and has been extensively implemented in intelligent residences, intelligent cities, and healthcare systems. Smart devices connected to Wi-Fi networks can be controlled by people.

The environmental parameters WSNs can track are temperature, humidity, sound, light, smoke, gas, etc. WSNs are used in the identification of forest fires, forest flow prediction, etc. In our everyday lives, HetIoT applications must also have user-friendly interfaces to make the applications easier to use. HetIoT uses secure and reliable smart devices to move data via satellites to the cloud server layer. Based on analytical data results the cloud servers will monitor the devices remotely.

3.5.4.1 Applications of Heterogeneous Network for IoT

Since 1999, IoT has progressed at a rapid pace and has penetrated several industry sectors as per domestic and international data. The heterogeneous network for IoT is expected to be embraced by growing numbers of industries. Corresponding network constructions are available in various application fields. Social mobile networks can be developed using cellular phones, and indoor wireless security networks are commonly used. But, because of their long-range transmitting capability, Wireless networks operate outside. Heterogeneous IoT network implementations are categorized because of application scenarios like industrial automation, precision farming, smart transport, smart home, and community safety.

3.5.5 SMART INDUSTRIAL

In industrial development, various smart devices were used. HetIoT applications have been implemented in supply chain management to purchase, inventory, sales, and other raw materials in companies improving the performance of the supply chain and reducing costs through optimization of the stream chain management systems. The technology for detection and decision-making will be improved to serve decision-making for the stream chain network systems of companies [37]. In process optimization, the level of the production process lines, acquisitions of real-time parameters, monitoring of equipment, and control of material usage have been enhanced by HetIoT applications. For instance, for production management and maintenance in manufacturing, Brizzi et al. [38] suggested an industry model. The model can also obtain data on energy consumption in real-time. Industrial HetIoT focuses on device surveillance and energy management in monitoring applications. To confirm protection during the operation of equipment, we may monitor oil and gas stations remotely.

The basis for the Industrial HetIoT can be considered as Industry 4.0. To build an adaptive, energy-efficient intelligent factory, Industry 4.0 incorporates industry-related technologies. In accordance with [39], the reference architecture centered on

HetIoT's intelligent factory and identified the characteristics of the factories focusing on sustainable development. They also addressed the anticipated benefits by presenting a system focused on HetIoT's intelligent model energy plant management.

3.5.6 SMART AGRICULTURAL

Agricultural IoT products automatically enable or disable the devices on or off based on sensitive data such as soil conditions, soil temperature, CO_2 concentrations, humidity and light signals, leaf moisture, and other environmental parameters in real-time. The majority of the agricultural precision systems are WSNs, and energy use is a problem for sensing areas deprived of power supply [40]. In addition, topology creation is another field of study, which can increase cyber attack robustness. The environmental parameters are measured by WSNs, which comprise several sensors that have been installed in a surveillance area and are designed as autonomous and multiple hops. Cellular or WMN (i.e. 3G/4G/LTE/5G) sensing data are transmitted to the cloud server.

Users' needs mean that agricultural HetIoT is ready for processing anytime, that integrated ecological information is automatically monitored on agricultural installations, that the environment is controlled automatically, and that smart management is provided with the scientific basis to certify the most appropriate evolution environment for crops. Environmental information is gathered in real-time by sensor devices within the greenhouse and transmitted via a mobile communications network to the service management system. The data is analyzed and processed by the service management platform.

3.5.7 SMART HOME

A smart home denotes an effective management system for family matters that employ different technology, such as integrated wiring, network connectivity, protection, automatic control, audio, and video. The management system expands protection, comfort, art, and the climate. The management system improves home safety. Wi-Fi networks consist of wireless remote data networks (e.g. 3G/4G/LTE/5G) and short-range wireless link networks, connect smart devices with each other (i.e., Bluetooth, infrared, RFID). Furthermore, cellular networks are built by users and smart mobile devices, and users can access the smart home management system remotely.

Intelligent home products must respond to users' real needs and offer a spontaneous user interface. For instance, Sim et al. [41] have implemented sound-conscious technologies for user behaviors, to enhance the serviceability of the device. The smart home's intelligent subsystems must function 24 hours a day, attaching rank to protection, consistency, and device fault tolerance.

Smart home not only has conventional residential features, but it also offers complete connectivity between houses, appliances, and automation equipment as compared to a common home. New technology of energy-saving Wi-Fi in smart homes has been used. Li et al. [42] assessed the energy use and interrupt downlink

Simulation Platforms

efficiency of intelligent home systems. Results show that this technology increases not only smart grid capacity but also reliability in the home. Liu et al. [43] have suggested a model of research, integrating home smart grid and renewable energy systems, and evaluating the model, to identify the lowest energy costs and economic energy management schemes. On the other hand, Anvari-Moghaddam et al. [44] specify an algorithm for intelligent home energy optimization, considering the balance between energy saving importance and comfort. Not only does the algorithm save resources but also make residents comfortable.

3.5.8 Intelligent Transportation System

Intelligent transportation Ssystem (ITS) is the forthcoming Internet of vehicle networks for transport. Due to the mobility of vehicles, nodes can freely go and enter vehicle networks on the internet. The resolutions for fixed and mobile devices for exchanging data on vehicle networks are Bluetooth, Infrared, and RFID. Tacconi et al. [45] used WSNs to support ITS by efficiently incorporating state-of-the-art IT and electronic sensor technology into the ground transport system.

Modern information technology makes ITS possible, which makes it easier for commuters to gather, process, disseminate, share, and analyze information. Younes and Boukerche [46] suggested a protocol to identify congestion detecting areas of congested urban grid areas. To satisfy the application requirements, ITS may also use a number of networks. To ensure traffic safety and increase transport quality, existing transport facilities will efficiently use ITS to lessen traffic loads and environmental pollution. Vehicle ad-hoc networks (VANETs) emerge as an important technology for ITS, with ITS being widely adopted. Baiocchi et al. [47] suggested the TOME method for measuring VENET traffic using phones. In tens of seconds, TOME can collect information on traffic in real-time and provide an accurate estimate.

3.5.9 Smart Healthcare

In several areas of healthcare, IoT technology penetrated from patient vital sign monitoring to monitoring and instruction recovery exercises to individual everyday activity tracking [48] and operating rooms [49]. Healthcare has encouraged the production and reorientation of wearable smart devices. Today, smartphones, smartwatches, intelligent bracelets, smart head-mounted devices, and other wearables detect people's cardiac velocity, blood pressure, sleep condition, and activities. These sensing data can provide the users with their own health analyses and health recommendations. Moreover, it is possible to prevent certain occult diseases. The physicians can monitor patients' physical condition in real-time from a distance through medical IoT technology [50].

HetIoT should be person-oriented and intelligent machines should make the surgery safer for patients and physicians. The automated medical IoT method has been accelerated by hospital management systems. Precision medicine is the future of healthcare by analyzing sensing data from patients. In addition, IoT and cloud computing will support the entire medical industry [51,52–54].

3.6 SUMMARY

This chapter addresses the various design issues of wireless sensor networks, applications with topology control and energy models. Different simulator models addressed their advantages and drawbacks as well.

Exercise
1. List and explain different topology control design issues.
2. Explain topology awareness problem.
3. List types of topology control problems. Explain each type of topology control in brief.
4. Enlist and explain the different types of network models.
5. Compare the different types of network models.
6. Explain the MATLAB simulation model for IoT domain.
7. Describe future research direction: heterogeneity of network technologies.

REFERENCES

[1] Karp, B. and Kung, H. T. 2000. Greedy Perimeter Stateless Routing for Wireless Networks. In proceedings of the Sixth Annual International Conference on Mobile Computing and Networking (Mobicom).
[2] Bulusu, N., Heidemann, J. and Estrin, D. 2000. GPS-less Low Cost Outdoor Localization for Very Small Devices. IEEE Personal Communications Magazine.
[3] Liu, Y., Xiao, L., Liu, X., Ni, L. M. and Zhang, X. 2005. Location awareness in unstructured peer-to-peer systems. *IEEE Transactions on Parallel and Distributed Systems (TPDS)*, 16:February, 163–174.
[4] Kranakis, E., Singh, H. and Urrutia, J. 1999. Compass Routing on Geometric Networks, in proceedings of the 11th Canadian Conference on Computational Geometry.
[5] Bose, P., Morin, P., Stojmenovic, I. and Urrutia, J. 2001. Routing with guaranteed delivery in ad hoc wireless networks. *Wireless Networks*, 7:609–616.
[6] Kuhn, F., Wattenhofer, R., Zhong, Y. and Zollinger, A. 2003. Geometric Ad-Hoc Routing: Of Theory and Practice. In proceedings of ACM PODC.
[7] Douglas, S., Couto, D. and Morris, R. 2001. Location proxies and intermediate node forwarding for practical geographic forwarding. MIT Laboratory for Computer Science MITLCS-TR-824.
[8] Li, Q. and Rus, D. 2000. Sending Messages to Mobile Users in Disconnected Ad-Hoc Wireless Networks. In proceedings of ACM Mobicom.
[9] Yu, Y., Govindan, R. and Estrin, D. 2001. Geographical and Energy Aware Routing: A Recursive Data Dissemination Protocol for Wireless Sensor Networks. UCLA Computer Science Department UCLA/CSD-TR-01–0023.
[10] Fang, Q., Gao, J. and Guibas, L. J. 2004. Locating and Bypassing Routing Holes in Sensor Networks. in proceedings of IEEE INFOCOM.
[11] Li, M. and Liu, Y. 2004. Wireless Sensor Network for Underground Monitoring. submitted to ACM Sensys'06.
[12] Wood, D., Stankovic, J. A. and Son, S. H. 2003. JAM: A Jammed-Area Mapping Service for Sensor Networks. In proceedings of 24th IEEE Real Time System Symposium (RTSS).
[13] Wood, D. and Stankovic, J. A. 2002. Denial of service in sensor networks. *IEEE Computer Magazine*, 35:48–56.

[14] Hu, Y. C., Perrig, A. and Johnson, D. B. 2002. Wormhole Detection in Wireless Adhoc Networks. Department of Computer Science, Rice University, Tech Rep TR01-384, 200.
[15] Karlof and Wagner, D. 2003. Secure Routing in Wireless Sensor Networks: Attacks and Countermeasures. in proceedings of 1st IEEE International Workshop SNPA.
[16] Khan, Muhammad Asghar, Khan, Asfandyar, Shah, Said Khalid and Abdullah, Azween 2012. An Energy Efficient Color Based Topology Control Algorithm for Wireless Sensor Networks. Institute of Engineering and Computing Sciences, University of Science and Technology Bannu, Bannu, Pakistan.
[17] Ye, F., Zhong, G., Lu, S. and Zhang, L. 2003. PEAS: A Robust Energy Conserving Protocol for Long-lived Sensor Networks. in proceedings of International Conference on Distributed Computing Systems (ICDCS).
[18] Cao, Q., Abdelzaher, T., He, T. and Stankovic, J. 2005. Towards Optimal Sleep Scheduling in Sensor Networks for RareEvent Detection. in proceedings of IPSN.
[19] Zhang, H. and Hou, J. 2003. Maintaining Sensing Coverage and Connectivity in Large Sensor Networks. Department of Computer Science, UIUC UIUCDCS-R-2003–2351.
[20] Wang, X., Xing, G., Zhang, Y., Lu, C., Pless, R., et al. 2003. Integrated Coverage and Connectivity Configuration in Wireless Sensor Networks. In proceedings of ACM SenSys.
[21] Huang, F. and Tseng, Y. C. 2003. The Coverage Problem in a Wireless Sensor Network. In proceedings of ACM WSNA.
[22] Xue, W., Luo, Q., Chen, L. and Liu, Y. 2006. Contour Map Matching For Event Detection in Sensor Networks. In proceedings of ACM SIGMOD.
[23] Wang, G., Cao, G. and Porta, T. L. 2004. Movement-Assisted Sensor Deployment. In proceedings of IEEE INFOCOM.
[24] Aurenhammer, F. 1991. Voronoi Diagrams - A Survey of a Fundamental Geometric Data Structure. ACM Computing Surveys.
[25] Du, F. Hwang and Fortune, S. 1992. Computing in Euclidean Geometry. Edited by Ding-Zhu Du and Frank Hwang, *World Scientific, Lecture Notes Series on Computing*, Vol. 1.
[26] Heo, N. and Varshney, P. K. 2003. An Intelligent Deployment and Clustering Algorithm for a Distributed Mobile Sensor Network. in proceedings of IEEE International Conference on Systems, Man and Cybernetics.
[27] Howard, Mataric, M. J. and Sukhatme, G. S. 2002. Mobile Sensor Network Deployment using Potential Fields: A Distributed, Scalable Solution to the Area Coverage Problem.
[28] Batalin, M. A. and Sukhtame, G. S. 2004. Coverage, exploration and deployment by a mobile robot and communication network. *Telecommunication Systems Journal, Special Issue on Wireless Sensor Networks*, 26(2):181–196.
[29] Wang, G., Cao, G. and Porta, T. L. 2003. A Bidding Protocol for Deploying Mobile Sensors. In proceedings of IEEE International Conference on Network Protocol (ICNP).
[30] Narayanaswamy, S., Kawadia, V., Sreenivas, R. and Kumar, P. 2002. Power control in ad hoc networks: Theory, architecture, algorithm and implementation of the COMPOW protocol. In proceedings of European Wireless.
[31] Kirousis, L., Kranakis, E., Krizanc, D. and Pelc, A. 2000. Power consumption in packet radio networks. *Theoretical Computer Science*, pp. 289–305. 10.1007/BFb0023473
[32] Xu, Y., Heidemann, J. and Estrin, D. 2001. Geography-informed Energy Conservation for Ad Hoc Routing. In proceedings of ACM Mobicom.

[33] Chen, B., Jamieson, K., Balakrishnan, H. and Morris, R. 2001. Span: An Energy-Efficient Coordination Algorithm for Topology Maintenance in Ad Hoc Wireless Networks. In proceedings of Mobicom.

[34] Schurgers, C., Tsiatsis, V., Ganeriwal, S. and Srivastava, M. 2002. Topology Management for Sensor Networks: Exploiting Latency and Density. In proceedings of ACM Mobihoc.

[35] Qiu, T., Chen, N., Li, K., Qiao, D. and Fu, Z. 2016. Heterogeneous ad hoc networks: Architectures, advances and challenges. *Ad Hoc Network*, 55:143–152, Feb.

[36] Jo, M., Maksymyuk, T., Strykhalyuk, B. and Cho, C.-H. 2015. Deviceto-device-based heterogeneous radio access network architecture for mobile cloud computing. *IEEE Wireless Communications*, 22(3):50–58, Jun.

[37] Li, J., Cheng, X. and Liu, B. 2013. Research on complex event of Internet of Things for supply chain decision support. *ICIC Express Letters B Applied International Journal of Research Surveys*, 4(5):1481–1487.

[38] Brizzi, P. et al. 2013. Bringing the Internet of Things along the manufacturing line: A case study in controlling industrial robot and monitoring energy consumption remotely. In Proceedings IEEE 18th Conference on Emerging Technology Factory Automation (ETFA), pp. 1–8.

[39] Shrouf, F., Ordieres, J. and Miragliotta, G. 2014. Smart factories in industry 4.0: A review of the concept and of energy management approached in production based on the Internet of Things paradigm. in Proceedings IEEE International Conference on Industrial Engineering and Management, pp. 697–701.

[40] Patota, F. et al. 2016. DAFNES: A distributed algorithm for network energy saving based on stress-centrality. *Computer Network*, 94:263–284.

[41] Sim, J. M., Lee, Y. and Kwon, O. 2015. Acoustic sensor based recognition of human activity in everyday life for smart home services. *International Journal of Distributed Sensor Networks*, 11(9):1–24, Jan.

[42] Li, Z., Liang, Q. and Cheng, X. 2014. Emerging WiFi direct technique in home area networks for smart grid: Power consumption and outage performance. *Ad Hoc Network*, 22:61–68, Nov.

[43] Liu, G.-R., Lin, P., Fang, Y. and Lin, Y.-B. 2015. Optimal threshold policy for in-home smart grid with renewable generation integration. *IEEE Transactions Parallel Distributed Systems*, 26(4):1096–1105, Apr.

[44] Anvari-Moghaddam, A., Monsef, H. and Rahimi-Kian, A. 2015. Optimal smart home energy management considering energy saving and a comfortable lifestyle. *IEEE Transactions Smart Grid*, 6(1):324–332, Jan.

[45] Tacconi, D., Miorandi, D., Carreras, I., Chiti, F. and Fantacci, R. 2010. Using wireless sensor networks to support intelligent transportation systems. *Ad Hoc Network*, 8(5):462–473.

[46] Younes, M. B. and Boukerche, A. 2015. A performance evaluation of an efficient traffic congestion detection protocol (ECODE) for intelligent transportation systems. *Ad Hoc Network*, 24:317–336, Jan.

[47] Baiocchi, A., Cuomo, F., De Felice, M. and Fusco, G. 2015. Vehicular adhoc networks sampling protocols for traffic monitoring and incident detection in intelligent transportation systems. *Transportation Research C: Emerging Technologies*, 56:177–194, Jul.

[48] Lun, R., Gordon, C. and Zhao, W. 2016. The design and implementation of a Kinect-based framework for selective human activity tracking. in IEEE International Conference on Systems, Man and Cybernetics, Budapest, Hungary, Oct. pp. 2890–2895.

[49] Zhao, W. 2015. Towards trustworthy integrated clinical environments. In Proc. IEEE 12th Int. Conf. Auton. Trusted Comput., Aug., pp. 452–459.

[50] Bader, A., Ghazzai, H., Kadri, A. and Alouini, M.-S. 2016. Front-end intelligence for large-scale application-oriented Internet-of-Things. *IEEE Access*, 4:3257–3272.

[51] Hossain, M. S. and Muhammad, G. 2016. Cloud-assisted industrial Internet of Things (IIoT)-Enabled framework for health monitoring. *Computer Network*, 101:192–202, Jun.

[52] Qiu, Tie, Chen, Ning, Li, Keqiu, Atiquzzaman, Mohammed and Zhao, Wenbing 2018. How can heterogeneous Internet of Things build our future: A survey. *IEEE Communications Surveys & Tutorials*, 20(3): 2011–2027. 10.1109/COMST.2018.2803740

[53] Sengupta, Debasmita and Roy, Alak 2014. A Literature Survey of Topology Control and Its Related Issues in Wireless Sensor Networks. *International Journal of Information Technology and Computer Science*. 10.5815/ijitcs.2014.10.03

[54] Issariyakul, Teerawat 2012. Introduction to Network Simulator 2 (NS2). Introduction to Network Simulator NS2.

4 Link Efficiency-Based Topology Control Algorithm for IoT Domain Application

4.1 INTRODUCTION

The World's physical and computational world is associated with a WSN by smart devices known as sensor nodes or motes. An ad hoc network of wireless sensors includes small autonomous devices that can sense its environment within a certain metric space, and communicate with local computations via radio transmissions. In energy-restricted wireless sensor node networks, the main challenge is to maximize network life. Some nodes can quickly exhaust their energy or become dysfunctional in the process. Topology monitoring is an important method for enhancing the energy efficiency of wireless sensor networks. As energy-efficient topology control algorithms developed one of the critical design parameters, the focus is to develop efficient topology control algorithms to expand the network performance. Hence, a new topology control scheme based on fairness, admissibility, and effectiveness of nodes using RSSI is used. The dynamics of removing unnecessary connections in the dense network are addressed. It increases the life of the sensor nodes while preserving connectivity and guarantees an effective energy connection to the network.

4.1.1 Received Signal Strength Indicator

The energy of receiving signal is measured by RSSI. In the case of radio frequency waves, power is inversely proportional to distance. The energy of transmitting a signal can be measured and distances d calculated. The power of receiving signal is calculated by subtracting path loss of transmitted power. RSSI stands for received signal strength indication, which indicates the power of a signal on any wireless or radio link. Its unit is dB. RSSI is used for receiving the end of the channel. RSSI is used for the estimation of node connectivity and node distance. RSSI is used for sampling the channel power. The relationship between distance and RSSI values is the foundation for wireless sensor networks represented in Equation (4.1) [1].

$$P_r = P_t \cdot (1/d)^n \qquad (4.1)$$

This relationship becomes the key to positioning technologies and ranging in wireless sensor networks represented in Equation (4.2).

$$10 lg P_r = 10 lg P_t - 10\, n\, lg d \tag{4.2}$$

The connection between distance and RSSI values can be better described by the log-normal shadowing model (LNSM). LNSM is a signal propagation model [2]. The variations of RSSI values along with distance changes by evaluating a large experimental database. The association function of distance and variance of RSSI is based on the outcome of the analysis and the log-normal shadowing model with dynamic variance is established. Ranging technologies most commonly used are RSSI ranging, TOA, TDOA, and AOA. Implementation complexity, cost, and communication overhead of RSSI ranging are very low as compared to other technologies. Because of these properties, RSSI ranging is most suitable for limited power nodes in wireless sensor networks.

The association between the distance between nodes and transmitted/power of the wireless signal is described by RSSI ranging principle and shown in (4.3).

$$P_R dBm = A - 10 n lg\, d \tag{4.3}$$

The received power of the wireless signal is shown by the Pr. The transmitted power of the wireless signal is shown by Pt., the distance between sending and receiving node is shown by d and the transmission factor is shown by n. The value of the transmission factor depends on the propagation environment.

4.1.2 Limitation of RSSI

The independent hardware interface layer in the standard platform does not provide RSSI data. RSSI values provided by TinyOS are not in the form of dBm units, so the platform-specific layer is responsible for converting it into meaningful data. In cell phones, RSSI is usually in the form of several bars. Signal measurement in the industry is in the form of negative dBm. The signal −80 dBm or less to provide good quality of the signal. −62 dBm signal quality is better than −80 dBm signal. If the signal has greater than −80 dBm value, then no call dropping, no degradation in voice quality. If a signal value is −90 dBm to 100 dBm then a high risk of call dropping and reduction in voice quality. If a signal value is less than 110 dBm, then it provides no link at all.

4.2 NETWORK MODEL

The sensor system is modeled as a directed graph G(V, E) where V' is an arrangement of nodes and E' is an arrangement of links [3]. Gdiagraph represents the diagraph of the system with Ediagraph represents the arrangement of undirected edges. For every node, i have a one-of-a-kind identity id. Id is spoken to in the Euclidian plane with its directions. A directed edge between two nodes i and j is

Algorithm for IoT Domain Application

indicated as [i->j], [i->j] € Ediagraph and has a separation of d(i,j). The set of neighbors of i with which i is specifically associated are signified as the set N(i) and characterized as N(i): [i->j] Ediagraph. Let N(i) contains the identity, energy saves, eligibility parameter, and required transmission energy to achieve every neighbor. The transmission power from node i to node j is signified as Pt – it. All nodes begin with equivalent introductory battery limit E. Deployment of nodes in a rectangular field of arrangement and organization is displayed by utilizing a Poisson point process. The purposes of the Poisson process are similarly prone to happen inside the limit of A. We can discover the likelihood of n nodes is given in Equation (4.4)

$$Pr[n \text{ nodes in } A] = e^{-m \cdot} (m. A)^2/n!. \qquad (4.4)$$

4.2.1 Definitions

Definition 1: All the neighbors of i are denoted as

$$N(i): [i \leftrightarrow j] \in E_{diagraph}$$

Definition 2: Received signal power can be defined as

$$Pr_x \alpha \; P_{tx}/d^Y \text{ where } Y = \text{Path loss exponent.}$$

The wireless medium is used for communication. Where λ is the density.

Definition 3: A graph is k-node associated if there does not exist a set of k −1 vertices whose exclusion disconnects the graph. Let V' V with 'V'| < k be the set of vertices that should be removed. Then for the graph

$$G' = (V', E') = =(V\backslash V', E\backslash (V' \times V')) \text{ we have } \nexists x, y \in V \text{ in } G'.$$

Defination 4: Network rehabilitees $\lambda(G)$ is defined as

$$\lambda(G) = E(|Vr|)/|V|$$

Referred these definitions as [3],

$$K = \frac{D}{\alpha \cdot dchar} = \frac{Kopt}{\alpha}$$

$$P_{relay}(d) = (\alpha_1 + \alpha_2 d^\lambda) r$$

$$P_{link-min}(D) = K'_{opt} \cdot P_{relay}(d_{char})$$

$$P_{link}(D') = \sum_{i=1}^{k} Prelay(ci \cdot dchar)$$

$$\frac{Plink - min(D)}{Plink(D')} = \frac{K'_{opt} \cdot Prelay(dchar)}{\sum_{i=0}^{n} Prelay(ci \cdot dchar)}$$

4.2.2 Assumption

Table 4.1 describes various parameters used for link-based efficient topology control protocol for sensor networks.

4.3 IMPROVED LINK EFFICIENCY-BASED TOPOLOGY CONTROL ALGORITHM

The basic model for the protocol for topology control is established [3]. For any arbitrary node in the network, the model determines the weighted relating area in

TABLE 4.1
Parameters and description

Sr. No	Parameter	Description
1	A	Bounded region
2	m	Poisson process density and it is correlated to the density of the network
3	N(i)	It is the group of i's neighbors to which it is related directly.
4	NL(i)	It is the table neighbor list and it stores set of each i.
5	P_{max}	It is the maximum transmission power. We have assigned different transmission power corresponds to respective neighboring node.
6	P_{ij}	The transmission power from node i to j.
7	E	Initial battery capacity.
8	d_{ch}	It is the characteristic distance (Y=2, d_{ch} is 100m/ Y=4, d_{ch} is 71 m)
9	Rt	Relay rate.
10	X_{11}	Transmitter electronics energy
11	X_{12}	Receiver electronics energy
12	Y	Path loss exponent (2 or 4)
13	X_2	Radio amplifier energy
14	L	System loss factor
15	I_t	Transmitting antenna gain
16	G_r	Receive antenna gain
17	H_t	Transmitting antenna height
18	H_r	Receiver antenna height

Algorithm for IoT Domain Application

two dimensions. The weighted area indicates the adjacent node's eligibility. Only local information is taken into account in the eligibility of each node.

4.3.1 Proposed Algorithm: LEBTC

The proposed algorithm consists of two phases, which are summarized as follows:

Phase 1:
INPUT:

    Nodes *NODES*{State*S*, Energy *E*},

    Threshold *THRESHOLD*,

    Energy consumption for each active step E_{ACTV},

    Energy consumption for each idle step E_{IDLE}

RESULT:

    Link Reduced Network

BEGIN

    NEIGHBOR_TABLE_DISTANCE = []
    NEIGHBOR_TABLE_ENERGY = []

    For each *NODE* **in** *NODES* **do**

        ENERGY = **CALCULATE-ENERGY***(N, E_{ACTV}, E_{IDLE})*
        RSSI= **CALCULATE-RSSI-METRIC***(N)*
        NEIGHBORS = **FIND-NEIGHBORS***(NODES, N)*
        DISTANCES= **CALCULATE-DISTANCE***(N, NEIGHBORS)*
        UPDATE-NEIGHBOR-TABLES*(N, RSSI, ENERGY, DISTANCES)*
    End for
    NEIGHBOR_TABLE_ENERGY = **REV_SORT***(NEIGHBOR_TABLE_ ENERGY)*
    NEIGHBOR_TABLE_ DISTANCE = **SORT***(NEIGHBOR_TABLE_ DISTANCE)*

    [ACTIVE_NODES, INACTIVE_NODES]
        = **APPLY-THRESHOLD***(THRESHOLD,*
    NEIGHBOR_TABLE_ENERGY,
        NEIGHBOR_TABLE_ DISTANCE)

    ACTIVATE*(ACTIVE_NODES)*
    SLEEP*(INACTIVE_NODES)*

END

Phase 2: The algorithm accomplishes a redundant process of edge removal without affecting connectivity in phase two. This process is considered to reduce the node degrees that enable interference reduction. If the node x has two neighboring nodes y, z ∈ N(x), such that the energy required to transfer from x to z directly is not less than the total energy to forward via y, we can eliminate w amongst neighbor list N (x).

4.3.2 MATHEMATICAL MODEL

Let S be the Whole System Consists:

$$S = \{V, E, P, G, F_{main} E_{diagraph}, Pr\}.$$

where
1. V is the set of all the network nodes.
2. E is the set of all connections between the network nodes.
3. P is a path function that sets the path for both nodes.
4. Let G is a graph,
5. $E_{diagraph}$: a set of undirected edges.
6. Pr: probability of n nodes

$$Pr[n \text{ nodes in } A] = e - m. (m. A)^2/n!.$$

Various definitions of network model are mentioned with mathematical representation:

- The free space propagation model:

$$Prx(d) = Ptx * Gtx * Grx * \lambda^2/(4\lambda)^2 * d^2 * L = Cf * Ptx/d^2$$

- The long-distance path model:

$$Prx(d) \alpha Ptx/D^\alpha$$

The path loss is proportional to the transmission power of *Ptx* to *d*.

7. F_{main}: Set of all the function's parameters used
   - [Nbr_list,location_arr]: Neighbor table is prepared for transmitting node.
   - nbr_src: source and information of nodes.
   - Function Ptx=calc_src_tx_power(s,Next_node): Distance D of all the neighbors of transmitting node is maintained in the table.
   - Node_info(Next_node).energy: Defining the Threshold value for an Energy E of a neighbor node.

The algorithm is designed for the above-described network model. The algorithm process in two phases with all the mathematical calculations. The first phase provides all the calculations and neighboring table. RSSI metric and energy are calculated. In the second phase, a duplicate edge removal procedure is performed without interfering with connectivity. This step aims to reduce the degrees of the node which contributes to interference reduction.

4.3.3 FLOW DIAGRAM

The working topology control algorithm and LEBTC data flow diagram are defined in Figures 4.1 and 4.2.

4.4 IMPLEMENTATIONS

MATLAB is a simulation tool that uses a technique defined by Lewis and shelter to randomly deploy node positions. The simulation environment assumes 100 nodes are deployed in a two-dimensional plane of 500 m × 500 m. The parameters used in the simulation are assumed in Table 4.2. Table 4.2 displays the implementation parameter.

The proposed algorithm requires an initial graph upon which the proposed algorithm has been applied (see Figures 4.3, 4.4, and 4.5).

4.4.1 RNG-RELATIVE NEIGHBORHOOD GRAPH

The relative neighbor graph (RNG) removes the longest edge of each that comprises two neighbors and itself. Formally, the RNG $G = (V,E)$ of a graph $G = (V,E)$ is defined as:

$$max\{d(x, z), \ d(y, z)\} < d(x, y)\}.$$

4.4.2 GG – GABRIEL GRAPH

The Gabriel Graph (GG) on the Euclidean plane, suggested by Gross and Yellen, 2004 [4]. It links point u and v if it has no other except itself and the neighbor that has a line segment uv as its diameter. Formally, the GG $G' = (V,E)$ of a graph $G = (V,E)$ is defined as:

$$d^2(x, z) + d^2(y, w) < d(u < v)$$

4.4.3 FETC AND FETCD

The author [3] focused on equality in a multi-Hop wireless sensor network and suggested a protocol on topology control that allows nodes to reasonably deplete their energy.

Connectivity is owned by the network and graph wireless sensor. One of the paths should be used at least to link the node to the base station in every node. The association between RSSI values and distance is the basis and the key of sensor

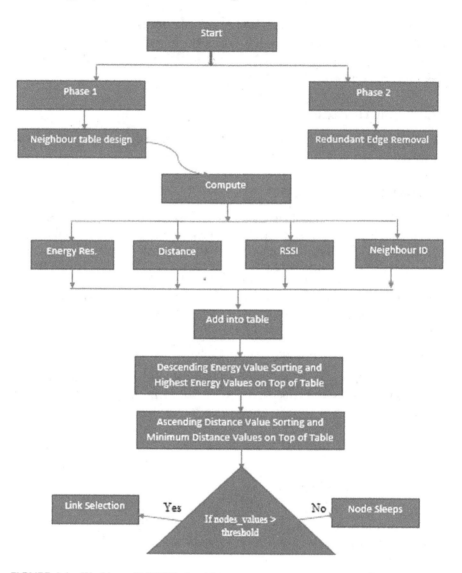

FIGURE 4.1 Working of LEBTC algorithm.

network range and location techniques. Log-normal shadowing model (LNSM) is an all-purpose signal propagation model [2]. The RSSI described the association between power communicated and wireless signal power and the distance between nodes [5]. P_r is the acknowledged power of wireless signals. Moreover, P_t is the communicated power of the wireless signal. d is the distance between the nodes that are sent and the nodes received. The transmission factor is considered as n. Figures 4.6 to 4.8 demonstrate the energy-aware routing for diverse iterations such as 2 iteration, 5 iteration, and 10 iterations respectively.

Figures 4.9 to 4.10 illustrate the shortest path routing for the same iteration i.e. ten times.

Algorithm for IoT Domain Application

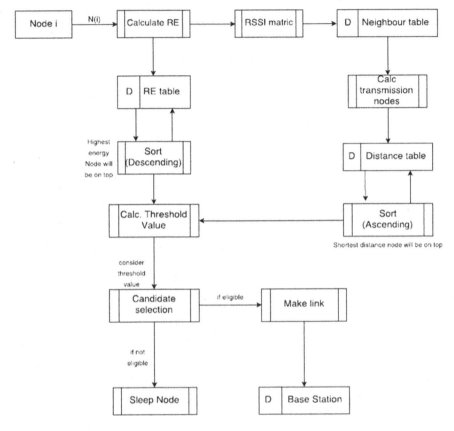

FIGURE 4.2 Data flow diagram for LEBTC.

TABLE 4.2
MATLAB parameter setting

Parameter	Value	Parameter	Value	Parameter	Value
Gt	1	Prx_thresh	85	alpha1	180 nJ/bit
Gr	1	Lamda	0.1224 m	alpha2	10 pJ/bit/m2 0:001 pJ/bit/m4
Ht	1.5 m	L	1	Alpha2	4
Hr	1.5 m	E	0	n_amp	0.023
Pt_max	0 dBm	alpha_11	26.5 mJ/bit	R	1

Figures 4.11 and 4.12 illustrate energy-aware routing after ten transmissions.

Figures 4.13 to 4.16 defines the various scenario for 100 and 300 nodes. Some sample cases are shown in the figure.

Figure 4.17 shows the link elimination after the topology control algorithm. LEBTC has improved the link elimination.

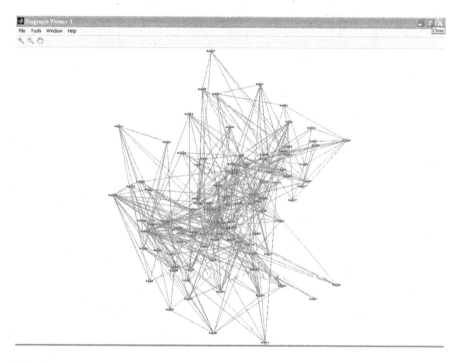

FIGURE 4.3 Topology formation using RNG.

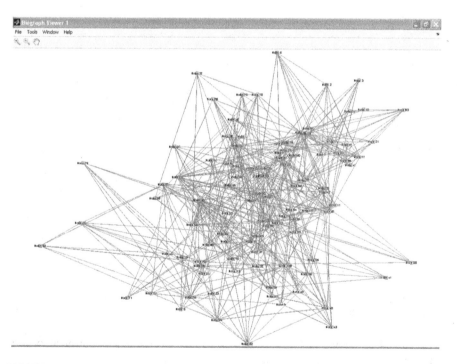

FIGURE 4.4 Topology formation using GG.

Algorithm for IoT Domain Application 103

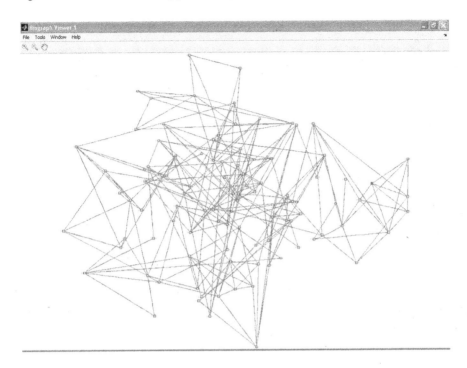

FIGURE 4.5 Topology formation using FETC.

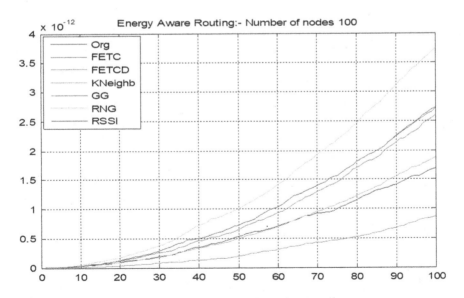

FIGURE 4.6 Energy aware routing (loop iterated two times).

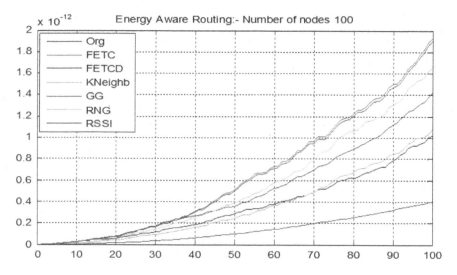

FIGURE 4.7 Energy aware routing (loop iterated five times).

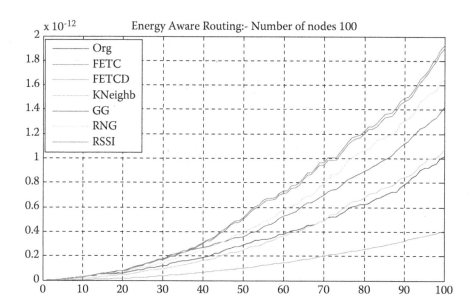

FIGURE 4.8 Energy aware routing (loop iterated ten times).

The comparative analysis for different time steps is defined in Figures 4.18 to 4.23. The results were measured and displayed in the bar graph at each point. In observing the plot, LEBTC concluded that the life of the sensor network is longer.

The performance of the best three algorithms per time steps for 100 nodes and 300 nodes are represented in Table 4.3 and Table 4.4. It is observed in the table for each of the steps, energy-aware routing, shortest path routing, and RSSI-based

Algorithm for IoT Domain Application

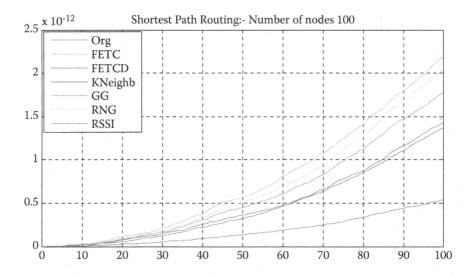

FIGURE 4.9 Shortest path routing (loop iterated ten times).

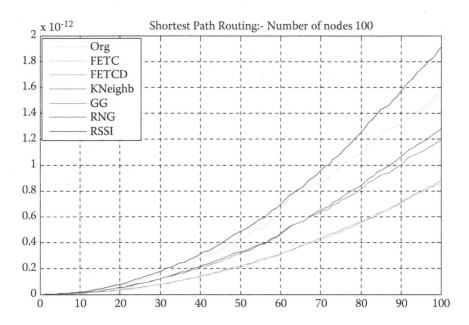

FIGURE 4.10 Shortest path routing (loop iterated ten times topology control period = after ten transmissions).

106 Energy Optimization Protocol Design for Sensor Networks in IoT Domains

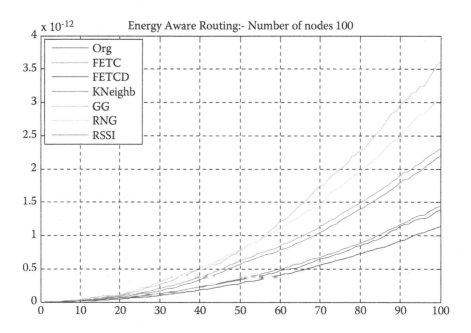

FIGURE 4.11 Energy-aware routing (loop iterated ten times topology control period = after ten transmissions).

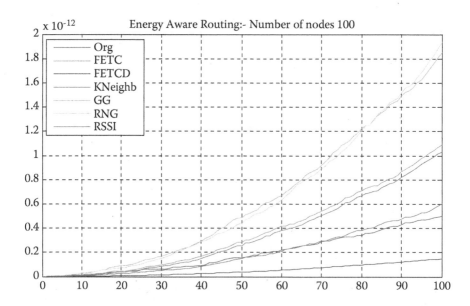

FIGURE 4.12 Energy-aware routing (loop iterated ten times topology control period = after five transmissions.

Algorithm for IoT Domain Application

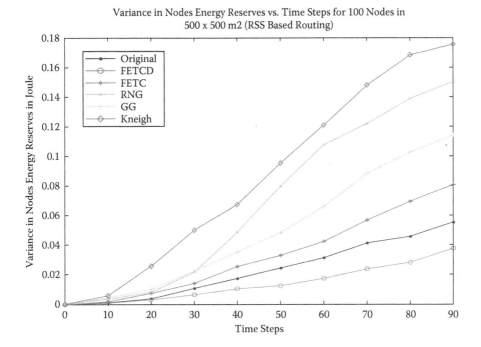

FIGURE 4.13 RSSI-based routing for 100 nodes.

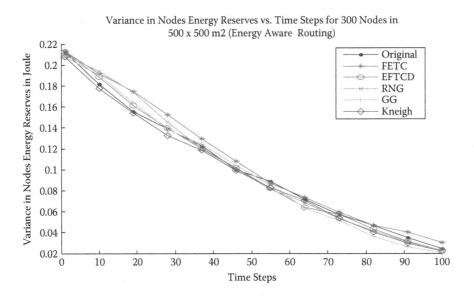

FIGURE 4.14 Energy-aware routing for 300 nodes.

108 Energy Optimization Protocol Design for Sensor Networks in IoT Domains

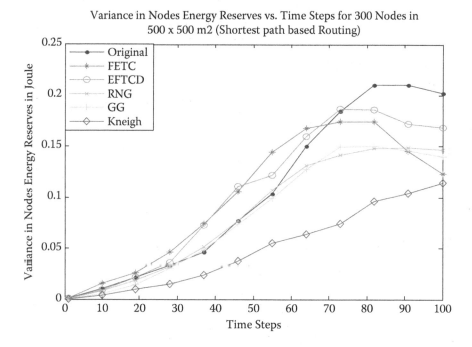

FIGURE 4.15 Shortest path-based routing for 300 nodes.

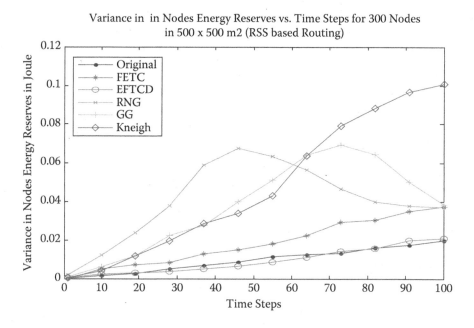

FIGURE 4.16 RSSI-based routing for 300 nodes.

Algorithm for IoT Domain Application

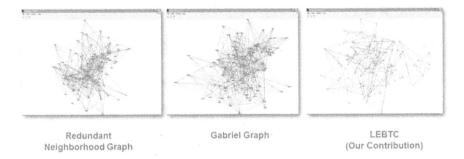

FIGURE 4.17 Comparison RNG, GG, and LEBTC.

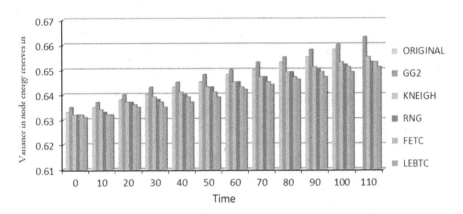

FIGURE 4.18 Performance and comparative analysis of energy-aware routing for 100 nodes.

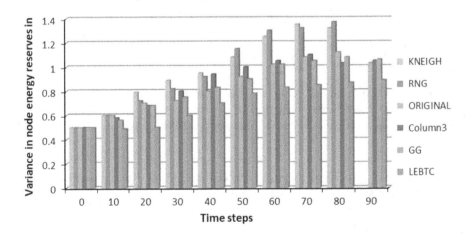

FIGURE 4.19 Performance and comparative analysis of shortest path routing for 100 nodes.

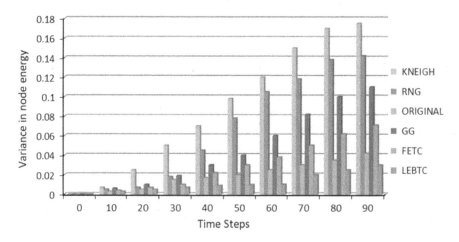

FIGURE 4.20 Performance and comparative analysis of RSSI-based routing for 100 nodes.

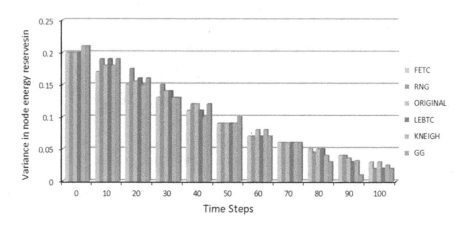

FIGURE 4.21 Performance and comparative analysis of energy-aware routing for 300 nodes.

routing were compared and best performed. Taking into account 90 sample simulation results, it is observed that LEBTC performed better as compared to RNG, GG, FETC, etc. Figures 4.24 and 4.25 demonstrate the top performers for 100 and 300 nodes.

4.5 FUTURE RESEARCH DIRECTION: GATEWAY PLACEMENT AND ENERGY-EFFICIENT SCHEDULING IN IoT

4.5.1 Overview

In the next decade, the Internet of things (IoT) will be one of the most disruptive innovations. The IoT paradigm [6] refers to the interconnection by Internet between different day-to-day physical devices, allowing them to communicate and share

Algorithm for IoT Domain Application

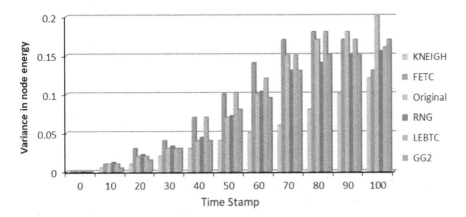

FIGURE 4.22 Performance and comparative analysis of shortest path routing for 300 nodes.

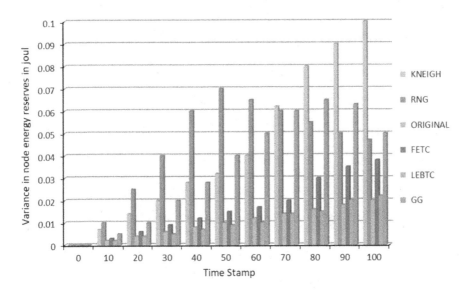

FIGURE 4.23 Performance and comparative analysis of RSSI-based routing for 300 nodes.

meaningful information. The quantity and complexity of the devices are expected to rise exponentially in the next few years. Some forecasts suggest that by the end of this decade we shall have between 50 and 100 billion IoT units [7].

Intel [8] and Microsoft [1] are several common IoT orientation architectures. A 3-layer structure, represented in Figure 4.26, is the common denominator of this design. IoT devices are in the bottom layer, gateway nodes, and aggregators in the middle layer, and cloud-based data centers are in the higher layer. The three layers may consist of various sublayers. For example, a few nodes in a hierarchy of gateway nodes serve as simple routers and some sophisticated analysis of different

TABLE 4.3
Percentage improvement over competitor algorithms of LEBTC (100 nodes)

Time steps	Position	0	10	20	30	40	50	60	70	80	90
Energy-aware routing for 100 nodes	1	LEBTC	LEBTC	LEBTC	LEBTC	LEBTC	LEBTC	LEBTC	LEBTC	LEBTC	LEBTC
	2	FETC	FETC	FETC	FETC	FETC	FETC	FETC	FETC	FETC	FETC
	3	RNG	RNG	RNG	RNG	RNG	RNG	RNG	RNG	RNG	RNG
Shortest path routing for 100 nodes	1	LEBTC	LEBTC	LEBTC	LEBTC	LEBTC	LEBTC	LEBTC	LEBTC	LEBTC	LEBTC
	2	FETC	GG	GG	Original	Original	GG	GG	GG	FETC	Original
	3	RNG	FETC	FETC	GG	GG	Original	Original	FETC	GG	FETC
RSSI-based routing for 100 nodes	1		LEBTC	LEBTC	LEBTC	LEBTC	LEBTC	LEBTC	LEBTC	LEBTC	LEBTC
	2		Original	Original	FETC	Original	Original	Original	Original	Original	Original
	3		FETC	FETC	Original	FETC	FETC	FETC	FETC	FETC	FETC

TABLE 4.4
Percentage improvement over competitor algorithms of LEBTC (300 nodes)

Time Steps	Position	0	10	20	30	40	50	60	70	80	90	100
Energy-aware routing for 300 nodes	1	LEBTC	FETC	FETC	FETC	KNEIGH	LEBTC	FETC	LEBTC	GG	LEBTC	GG
	2	FETC	Original	KNEIGH	KNEIGH	LEBTC	FETC	LEBTC	FETC	KNEIGH	GG	LEBTC
	3	RNG	KNEIGH	Original	GG	FETC	RNG	RNG	RNG	RNG	KNEIGH	RNG
Shortest path routing for 300 nodes	1		GG	GG	KNEIGH	KNEIGH	KNEIGH	KNEIGH	KNEIGH	KNEIGH	KNEIGH	KNEIGH
	2		KNEIGH	KNEIGH	LEBTC	GG	Original	GG	GG	RNG	RNG	FETC
	3		LEBTC	LEBTC	GG	Original	RNG	Original	Original	GG	GG	RNG
RSSI-based routing for 300 nodes	1		RNG	LEBTC	LEBTC	LEBTC	LEBTC	LEBTC	LEBTC	LEBTC	Original	Original
	2		LEBTC	Original	Original	Original	Original	Original	Original	Original	LEBTC	LEBTC
	3		Original	FETC	FETC	FETC	FETC	FETC	FETC	FETC	FETC	FETC

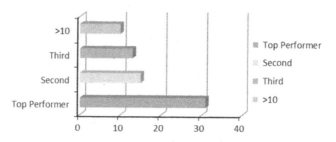

FIGURE 4.24 Top performer amongst link-based algorithms (100 nodes).

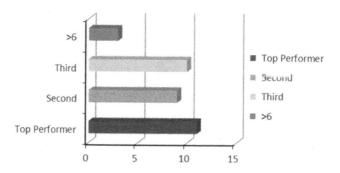

FIGURE 4.25 Top performer amongst link-based algorithms (300 nodes).

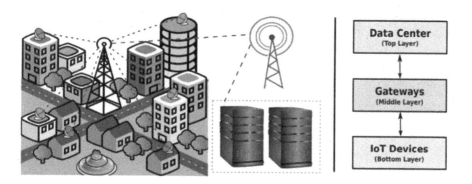

FIGURE 4.26 IoT architecture.

nodes. In the highest layer, there are multiple sublayers: client-side servers, cloud nodes, and storage nodes.

Energy consumption in compound IoT networks, as reported in the previous study [9,10], is seen as an equally significant problem. The details are that the highest IoT sensors and actuators are lightweight and very powerful. Typically, they use intermittent batteries or power sources such as solar power. In addition, there are no reliable energy sources at IoT hubs and gateways and are mainly located

Algorithm for IoT Domain Application

where energy supply alleviates the issue. Furthermore, given that IoT nodes normally manage several data, we need a great deal of energy to process and filter all data for processing in the future. Because of these limitations, we believe the goal of reducing energy use in IoT networks is worthwhile.

Figure 4.26 shows the IoT architecture. The IoT architecture of three layers includes specifically the subsequent layers:

1. **Bottom layer:** The bottom layer comprises wireless sensor networks (WSN) with different sensor nodes that can be used in various devices, for example, motors, smart homes, buildings, street lights, and traffic lights, for wearables, smartphones, shoes, and smart cards.
2. **Middle layer:** Gateway nodes that collect and buffer sensor node data normally carry out various data analysis measurement and transmit it across several hops into a data center.
3. **Top layer:** In general, the server in the data centers consists of this layer. These servers process the data they receive through the gateway, store the data in storage nodes and guide the actuators to adjust their environment. Messages to the actuators toward the gateways and hubs.

4.5.2 Placement of Gateways

Consider an architecture connected to each gateway by each sensor. As shown in Figure 4.27, the gateways are connected through a topology tree. The root gateway is the key to the servers (s). Task streams are produced by sensors and can be forwarded to the central server from a cloud server or a data center. The

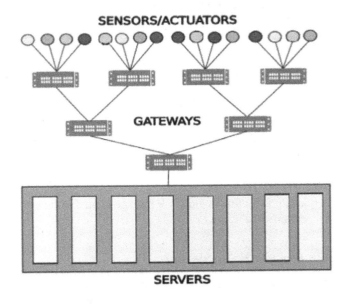

FIGURE 4.27 Architecture of an IoT system.

data were reached on the central server by a gateway (or hubs). Each gateway can also process and filter the data in its own right. Finally, the servers process the tasks and decide on the actions. This information is again informed through the same gateway network to the actuators (reverse-path). We believe that each node (sensor, gateway, or server) includes processing units that only store and forward or intelligently process packets. We aim to minimize energy overall without infringing on deadlines.

4.5.3 Task Model

A set of tasks *(t1, t2, t3,.)* generated by sensors is considered. A task t_i reaching a node can be modeled as a 4-tuples *(g_i, d_i, l_i, c_i)* operation, where g_i is the time to create a task, d_i is a deadline, l_i is the amount of network traffic related to a task in bytes and c_i is the number of running cycles needed to carry out the task.

4.5.4 Energy Consumption Model

Consider that any IoT processing node is an m-core device. Each core is compatible with r frequencies *(f1, f2, ..., fr)*, where *f1 < f2 < ... < fr*. The processor and the memory decide the energy consumption at a node. For such systems, this is a common assumption [11]. The CPU's energy has two components: dynamic and static, often referred to as leakage energy [12]. Becoming the primary contributor to our energy consumption, the dynamic component of our systems and temperature fluctuations are mainly focused on complex energy consumption. We think static energy is a constant [13].

Now, let n_j be the jth node on task t_i's path in the network. The energy consumption, e_{ij}, at node n_j is given by:

$$e_{ij} = \kappa_j \times n_{ij} \times f_j^2 \qquad (4.5)$$

where κ_j is a constant of proportionality (remains constant for a given node), n_{ij} is the number of cycles the task takes to execute at node j, and f_j is the execution frequency. The total energy is given by:

$$\begin{aligned} E_{total} &= \sum_{i=1}^{n} \sum_{j=1}^{l} e_{ij} \\ &= \sum_{i=1}^{n} \sum_{j=1}^{l} k_j \times n_{ij} \times f_j^2 \end{aligned} \qquad (4.6)$$

where n is the total number of tasks that have been generated, and l is the number of nodes in the execution path of the task.

4.5.5 Energy-Efficient Scheduling Algorithms

Two algorithms, global algorithm and local algorithm can be efficiently managed in the IoT network to ensure a minimum of delays have been breached. The time is

Algorithm for IoT Domain Application

taken for entering the actuator for each processing node to be determined for both algorithms when the current node leaves to measure the t_{rem} is used t_{est}. The current task is the maximum time available for the node and the time limit is taken into consideration. The DVFS (dynamic voltage frequency scaling) is then implemented by the multi-core node.

Each node tracks the average waiting times at the core and the frequency of different. When a task is planned, the availability of an idle core will first be verified. The task is scheduled if a core idle is available and the frequency of the core to the lowest available frequency greater than c/t_{rem} is set.

However, if the core is not idle, it is decided that an optimum frequency is used to minimize power consumption during a time of stay. Given that a discreet frequency range has the same or greater frequency as the optimal, we consider that frequency to be the least (f). The frequencies that operate each core are then taken into account. We restrict our search to those cores that have frequencies greater or equal to f and find the core with the lowest frequency in this range. This core operation is then presented to the EDF (earliest deadline first, a priority queue). The work would then take place on a time-limited basis.

During this method, the estimated time and test information is the most important parameter for DVFS performance. Therefore, two methods that make this information available to a node have been established.

4.5.5.1 Global Algorithm

We maintain a central, high bandwidth server, which is open to any node and allows all CS nodes to work and round around in the global algorithm. We have a CS (central server) In the system with several gateways linked to the Internet, this algorithm has more usefulness. We believe that all interaction with the CS has the highest priority. In [14] similar configurations were used [14].

Each node locally calculates t_{avg}, which means the average speed of performance of a task on the node by calculating the total of $(t_d - t_a)$, where t_a is the time the task arrives at the node, and t_d is the time the task leaves the node for all node tasks and divided it by the number of tasks. The time spent waiting for the task in the node's input queues and the time it takes to run at the core of the node includes this average runtime. Each node periodically sends this information every 100 microseconds about the average runtime.

This information is saved for each pair in a CS table (parent, child). This table generally saves the time required for a task to be performed in the parent node when it arrives from a specific child node. The parent can be given different kinds of tasks by different children. This is the t_{avg} value used by the CS to t_{est} any node. This is the value of t_{avg} that the CS will use to calculate t_{est} for any node. We expect the t_{est} to be stable in limited durations since we take steady streams of activities into account.

4.5.5.2 Local Algorithm

The individual node updates its t_{est} value with information from its neighboring nodes in the local algorithm. Any node calculates the average node time (t_{avg}) for a task to be performed by measuring the time interval $(t_d - t_a)$, where t_a is time, the task arrives at the node and t_d is the time when the node leaves the node to complete

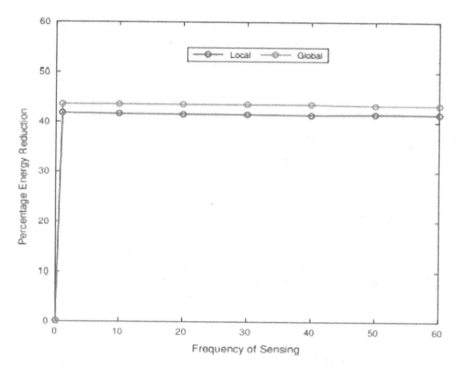

FIGURE 4.28 Energy consumption minimization (baseline: No-DVFS) v/s frequency of sensing.

all the tasks. The single-node sets the average node time. This interference includes the timeframe for waiting at the input queue of the node, assigning the EDF queue of the core, and performing it in the assigned core.

Moreover, every node keeps data regarding estimates of the time for each of its adjacent nodes (t_{est}). The estimated time of the node N represents the time that it takes for the actuator to reach the next node to which the operation is performed N after the current node. The t_{est} information that is used while the task is planned for a core corresponds to the next node where the task is performed. Before leaving a node, a message is backed up with the task. The message is a $t_{avg} + t_{est}$ value where the test value corresponds to the node from which the task was taken.

Furthermore, the local and the global minimum energy consumption of No-DVFS is shown in Figure 4.28 in an increment of 10 seconds with an average sensing time of 1 second and 10–60 seconds. The two algorithms also achieve a considerable energy reduction in heavy traffic scenarios. The efficiency of the algorithms is almost constant in terms of energy consumption when it comes to frequency sensing.

4.6 SUMMARY

In this chapter, the features of RSSI as a link efficiency metric have been discussed and a new link quality estimating measure is designed based on it. We have studied and implemented an existing protocol to control topology, such as RNG, GG, KNeigh,

FETC, etc. to improve on LEBTC by minimizing imbalance. Finally, gateway placement and energy-efficient scheduling in IoT are addressed in this chapter.

Exercise
1. What are the limitations of RSSI?
2. Explain the algorithm of LEBTC?
3. Compare following protocols with schematic diagram
   a. RNG
   b. GG
   c. FETC
4. Explain the future research direction of gateway placement and energy-efficient scheduling in IoT.

REFERENCES

[1] Microsoft. 2016. Microsoft Azure IoT Reference Architecture. http://download.microsoft.com/download/A/4/D/A4DAD253-BC21-41D3-B9D9-87D2AE6F0719/Microsoft_Azure_IoT_Reference_Architecture.pdf. (2016). Accessed on 8th December, 2017.

[2] Xu, J., Liu, W., Lang, F., Zhang, Y. and Wang, C. 2010. Distance measurement model based on RSSI in WSN. *Wireless Sensor Network*, 8:606–611.

[3] Dargie, W., Mochaourabb, R., Schill, A. and Guanc, L. August 2010. A topology control protocol based on eligibility and efficiency metrics. *The Journal of Systems and Software*, 84:1–10.

[4] Gross, J. L. and Yellen, J. (Eds.) 2004. *Handbook of Graph Theory*. CRC Press, Boca Raton.

[5] Dobircau, A., Folea, S., Bordencea, D. and Valean, H. 2012. System based on low-power Wi-fi technology for indoor localization of a mobile user, Proceedings of 2012 IEEE International Conference on Automation Quality and Testing Robotics.

[6] Sethi, P. and Sarangi, S. R. 2017. Internet of things: Architectures, protocols, and applications. *Journal of Electrical and Computer Engineering*, 2017 (2017):1–25.

[7] Sarangi, S. R., Goel, S. and Singh, B. 2018. Energy-efficient scheduling in IoT networks, Proceedings of the 33rd Annual ACM Symposium on Applied Computing – SAC'18.

[8] Intel. 2015. Intel IoT Platform Reference Architecture. https://www.intel.in/content/www/in/en/internet-of-things/white-papers/iot-platform-reference-architecture-paper.html. (2015). Accessed on 8th December 2017.

[9] Anastasi, G., Conti, M., Francesco, M. D. and Passarella, A. 2009. Energy conservation in wireless sensor networks: A survey. *Ad Hoc Networks*, 7, 3 (2009):537–568.

[10] Hammadi, A. and Mhamdi, L. 2014. A survey on architectures and energy efficiency in data center networks. *Computer Communications*, 40 (2014):1–21.

[11] Wang, J., Huang, C., He, K., Wang, X., Chen, X. and Qin, K. 2013. An energy-aware resource allocation heuristics for VM scheduling in the cloud. In HPCC_EUC. 587–594.

[12] Yan, L., Luo, J. and Jha, N. K. 2005. Joint dynamic voltage scaling and adaptive body biasing for heterogeneous distributed real-time embedded systems. *IEEE Transactions on Computer-Aided Design of Integrated Circuits and Systems*, 24, 7 (2005):1030–1041.

[13] Zhu, X., Yang, L. T., Chen, H., Wang, J., Yin, S. and Liu, X. 2014. Real-time tasks oriented energy-aware scheduling in virtualized clouds. *IEEE Transactions on Cloud Computing*, 2 (2014):1–14.

[14] Leung, N. K. N. and Hsu, R. T. 2005. Method and apparatus for out-of-band transmission of broadcast service option in a wireless communication system. (June 21 2005). US Patent 6,909,702.

5 Energy-Efficient Topology Control Algorithms for IoT Domain Applications

5.1 INTRODUCTION

Essential application in mission: packet loss is not sufficient. It is understood that packets in WSN are connected to their neighbors, packet loss is possible and thus reliable and energy efficiency should be improved. Topology control is a powerful approach for improving wireless sensor networks' energy productivity (WSNs). This is a beneficial but highly complicated technique. If it is not done carefully, an unwanted outcome can be achieved. The consideration of the topology control system during planning is important. Distributed algorithms, local data, local data requirements, connectivity, coverage, small node size, and ease. In the usual model, the model of the network is based on the suspicion of "connected" or "unrelated" certain nodes. The network is said to have a complete connection when all nodes are connected to the network. This technique is referred to as a topology control dependent on connectivity. Two phases of topology control are topology design and maintenance.

5.1.1 CONNECTED DOMINATING SET

The networks of wireless sensors are becoming more involved in several applications. Network management and lifetimes are the utmost important problems in WSN-based systems with their various characteristics and challenges. Connected dominating set (CDS) is recognized as an effective approach to managing the topology of networks, minimizing overheads and to increase the lifespan of networks.

Graph theory [1] provides the idea of a connected dominant set (CDS). It describes a collection of nodes for a particular graph associated. The CDS meaning is: For a specified connected graph (network) $G = (V, E)$, where V is a set of vertices (nodes) and E is the set of edges (the edge or connection between the two nodes is provided by any transmission range), the domino set (DS) is a subset of V' where, for respective vertex u of V, u is either a V' or at least one neighboring vertex of u is a V'. A DS is referred to as a CDS when it is bound to the subgraph caused by vertices in the DS.

Heuristics are categorized into two sets for CDS construction. The first generation of heuristics aims at finding a separate, maximum independent set (MIS) of nodes linked to the smallest tree or a Steiner tree. The next form is focused on developing a CDS through the cultivation of a tiny, trivial CDS.

DOI: 10.1201/9781003310549-5

The proposed design of two linked dominant set building algorithms, which provides better approximations. In this system, using a unit disk graph (UDG) for modeling a WSN in both approaches. The CDS construction was simplified by primary finding an autonomous set S_1 with the subsequent property: there are exactly three-hop distances amongst any two corresponding subclasses S_1 and S_2 of S_1. Secondly, the first step was to discover a minor set of S_2 nodes to dominate some of the detached components. In the second stage, the introduced S_2 nodes are connected to the S_1 nodes to form a dominant set, which returns the concluding CDS later adding additional relating nodes [2].

5.1.1.1 Approach-I

Approach-I builds a CDS through four principal stages. We have a coloring system in the CDS construction for illustration purposes to distinguish node states. There are labeled black S_1 (dominators). The nodes for the disconnected areas are colored red in Phase-2 (the set of nodes in S_2). The connectors are blue noticeable and the dominant nodes are grey noticeable. To facilitate the algorithm creation, other colors (white, orange, and yellow) are provisionally presented: white is used to initialize, orange to indicate nodes at some distance to a black, and yellow node to denote separated regions after S_1 building [2].

> Phase-1. Building S_1. The concept of independent sets guarantees a separation of at least two hops for each node pair in an independent set. The S_1 built in this step ensures that any complementary subsets have a hop distance of exactly three. In, we encompass the technique to build S_1 that meets the particular three-hop property. The construction scenario for S_1 is as follows: First, we set a tree-level u to the number of hops in T between u and I when I is the source of T, given a random root tree T spanning all nodes. Initially, entirely nodes are uncompressed and are highlighted in white. During the execution of the algorithm, nodes will be labeled with various colors. The root node I start with a black coloring of the S_1 structure. Then, it sends its 1-hop neighbors a *"BLACK"* post. When I get the message *"BLACK"* from a white surrounding, the white node deviates its color from white to grey and sends a message of *"GRAY"* to its one-hop neighbors. If the white neighbor accepts a *"GRAY"* note, the white node is changed from white to yellow and sends a message *"YELLOW"* from a white node, the white node itself orange. The color orange is provisional. Then, up until no orange or white nodes remain in the graph the algorithm repeats the following steps:
> 1. Choose an orange node u to mark it black. The orange node selected meets the two requirements: (a) level among orange nodes is the lowermost, and (b) it has 3-hop black neighbors at the highest point. The node u selected is black.
> 2. Node u transmits a message *"BLACK"* which is designed to overpower its 1-hop white, orange, and yellow nodes by marking them grey. A new grey node would then transmit a *"GRAY"* packet.

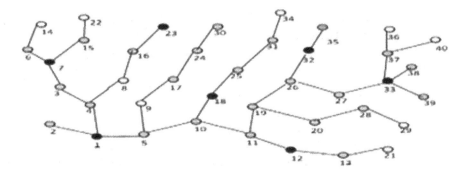

FIGURE 5.1 G after constructing S_1.

3. Upon receipt of a note entitled *"GRAY"* a white or orange node is yellow and transmits a note entitled *"YELLOW"*. Once received a *"YELLOW"* post, a white node mark itself as orange.

After the algorithm is done, any grey node will be certainly 1-hop from a black node that directs it and a somewhat yellow node will be two-hop from a black node; any nodes are black (S_1), grey (dominated), or yellow (detached components).

The following Figure 5.1, illustrates an instance of a graph G with 40 nodes initiated by node 1 following the S_1 construction. At least 2-hops from a black node is every yellow node. Yellow nodes, therefore, form several linked network graph components. These disconnected areas are seen to constitute a small part of the network. This connected part is handled in the algorithm mentioned in Phase 2.

Phase 2. **Disconnected areas coverage.** After Phase 1 is complete, the connected component types (yellow) are two: (a) the upright one with nodes on the same tree level, which are either 1-hop or are divided, and (b) the flat one which contains only certain leaf nodes in the spanning tree. In this step, a least dominant set is calculated in the following procedure for each connected variable. All the dominators calculated from this stage are red and form the S_2 set.

1. Let l be the lowermost node level for each vertically connected portion. Pick a yellow node, color red and grey from its neighbors if the element has only nodes in l. Therefore, the whole linked portion can be covered by a red node. If nodes on level $l + 1$ exist, choose the smallest amount of yellow nodes through l or $l + 1$ to cover all the yellow nodes on level l and the highest number of yellow nodes on level $l + 1$. Color these red nodes and all the grey neighbors yellow. Repeat this process until the linked component has no yellow node left.

2. The left-most node u and the color of the right-most Node v covering u and the supreme number of yellow nodes in the element start with each horizontal connected element. Color v

red and the grey of its neighbor. Reiterate this process up until the horizontally linked part leaves the yellow node.

Phase-3. S_2 nodes are linked to S_1 nodes. At least one black node, just two hops apart, exists in each red node. Therefore, one grey node must be included to bind a red node to its closest black node. However, a single connector may be used to link multiple red nodes to an S_1 node. It is preferable to have a limited number of connectors. So in this step, with a similar approach, we alter some grey nodes to blue. As a connector and a blue symbol, a grey node with the maximum number of red neighbors is used. If a grey node u is blue, all of its black/red one-hop neighbors are united in one area. At the end of this step, each connection-dominated portion is connected and combined with blue connectors to a region with at least one node.

Phase-4. Altogether, connecting S_1 Nodes. The fourth algorithm is used to bind every black node to the final CDS. At least a black neighbor is three-hop away from each black node. To link the nodes in S_1, at most 2 grey nodes have to be included to connect an S_1 node to the closest S_1 node. After this stage, the final CDS is obtained when connectors from S_1 to S_1 are found. The blue, red, and black union create the ultimate CDS.

5.1.1.2 Approach-II

There are four stages of the second method provided. Most of the phases of the approach-except phase-2 are identical.

Phase-1. Use the same algorithm used in Approach-I to construct S_1.

Phase-2. We have specified the yellow node x coverage factor as the numeral of its yellow surroundings. A gray/yellow x with the uppermost coverage factor is noticeable red and a post of *"RED"* is transmitted by x to monitor its gray-colored 1-hop yellow neighbors. All yellow nodes are colored red or grey when this process is completed.

Phase-3. We link CDC nodes to S_1 nodes at this point. If in Phase-2 of this approach a grey node u was marked red, u is already attached to a node S_1 and no U connectors are needed.

Phase-4. To locate the S_1-to-S_1 connectors by the same algorithm used in Approach-I of Phase 4.

More nodes will go to rest mode on the top of this system. 'Reliability of saving power trade.' The transmission of a viable topology monitoring convention is one of the main methods to drag the operational life of the sensor network. However, propose an intelligent clusters topology control estimate, providing energy-gathering nodes to improve the network's lifetime by promoting energy reinforcement in the sensor fields, valid base station location in the field such that a corresponding separation between group heads and base stations is minimized. In the light of overhead message, overhead energy, and reliability, our simulation results show that iPOLY consistently does better.

The convention frames a CDS [3,4] like a polyphonic network to achieve energy efficiency, giving reliability due to arbitrary disappointments in connecting. It adapts the remaining node energy to topological changes. The problem of extending the lifetime of the remote sensor network is widely sorted by the immediate methodology. Different expert calculations track the use of resources in aberrant methodology, while the rest of the technique provisions external support to extend network life. Although the aberrant method can encompass the life of the network, it does not discuss the impact of amplifying network life. Wireless sensor devices with truly high deceit rates are shoddy gadgets. Further, these gadgets must be thrown into the helicopter or comparable vertical excitement in many applications. Therefore, a few nodes interrupt or incompletely interrupt their utility. The stability of nodes is also affected by vital accessible energy levels.

Many gadgets to collect energy are proposed to add strength and power to the battery of the sensors. To increase obligation cycles and the planning of assignments to, for example, enhance the framework implementation, energy gathering has empowered networks to focus primarily on the issue of force administration to evaluate measures of energy that can be obtained afterward.

5.1.2 Clustering Mechanisms

Clustering means selecting node resources in the network and in the case of developing an efficient topology. For example, energy-saving, network density, or node recognition depends on different components for determining neighborhoods. An incredibly important priority of the group instrument over different systems like power adaptation or approaches to power mode is the development of an extremely flexible and directly supervisable topology with a different level structure. Another benefit of this extraordinary approach is the fact that sure errand can be restricted to a node setting called a group head and that packets from other non-bunch heads can be collected, prepared, and sent giving an efficient network association.

Other attractive features include load adjustment, information collection, and pressures on the information network for delayed life. Some methodologies change the option of group heads, which means that they are consumed relatively quickly because the tasks are intensive. The randomization of the procedure for determining cluster heads for distributing loads between many nodes on the network will effectively understand this small problem.

5.2 NETWORK MODEL

A few notes from the offered set are selected to form the virtual backbone within the created Topology Control Protocol. Give V and P the opportunity to organize nodes, $P \in V$ where every node in V is in P. Nodes are arranged in a hop next to various nodes.

$$p \in P(\forall \ v \in V \ \nabla \ \exists \ p \in V: (v, p) \in E)$$

Redundancy is the usual number of useful spreads in diagram G on trees. Each edge should be regarded as an extension of the tree spread. Network consistency is attained with the aid of tree propagation. To tackle irregular disappointments, there should be one crossing tree in the network. A graph G is indicated by an adjacency matrix is,

$$T = (Ti \cdot j) n \cdot n$$

Then

$$Ti, j = \begin{cases} 0 \\ 1 \end{cases}$$

A diagonal matrix is the degree of the vertices. If D indicates a graph G diagonal matrix then,

$$pi \cdot j = \begin{cases} \deg(vi) & for \ i = j \\ 0 & i \neq j \end{cases}$$

5.3 ENERGY-EFFICIENT ALGORITHM BASED ON CONNECTED DOMINATING SET

POLY's theoretical graphic control convention for WSN is distributed as a theoretical graph. The polygon in the network can be found by showing the network as a map. The convention framed a CDS like a polyphonic network to achieve energy efficiency and thereby offer reliability because of the arbitrary dissatisfaction of connections. It adapts the remaining node energy to topological changes. At the building level, connectivity must be preserved. The next step is the preservation of topology. The use of power in the remote sensor network is of central importance and shows a significant number of calculations, methods, and conventions for energy-saving use and lifetime development POLY is the convention on building topology. It depends on a polygon's probability. Due to the severe asset containment of the sensor nodes, Lifetime Expansion is a leading issue in the Wireless Sensor Network region [5,6].

Hassaan Khaliq Qureshi has the main credit for POLY's proposal. Our assumption is the same as the author's assumption in protocol use.

Topology construction protocol:
The topology building protocol is divided into various phases.

- Phase 1: In this step of the topological construction the Sink node initiates a neighboring discovery process which concludes the development of the CDS.
- Phase 2: Receipt of the next sink node list.
- Phase 3: This is the last stage of finding polygons in the graph.

5.3.1 Proposed Algorithm: iPOLY

INPUT:

Nodes *NODES* {State *S*, Energy *E*},

Energy consumption for each active step E_{ACTV},

Energy consumption for each idle step E_{IDLE}

RESULT:

Formed Polygon

BEGIN

CLUSTER_OF_NODES = []

BACKBONE_TABLE = []

CDS_TABLE = []

RANDOM-INITIALIZE-CDS *(NODES)';*

For each *NODE* **in** *NODES* **do**

ASSIGN-CLUSTER *(CLUSTER_OF_NODES, N)*

End for

BACKBONE_TABLE = **IDENTIFY-BACKBONE-MEMBERS** *(CLUSTER_OF_NODES);*

For each *BACKBONE_MEMBER* **in** *BACKBONE_TABLE* **do**

NEIGHBORS = **FIND-NEIGHBORS** *(NODES, N)*

End for

    INITAITE_POLYGON_FORMATION *(BACKBONE_TABLE)*

    TIMED_RECONSTRUCTION *(BACKBONE_TABLE)*

END

5.3.2 Mathematical Model

Let S be the Whole System Consists:

$$S = \{V, E, P, G, I_p, O_p,\}.$$

where

1. The network nodes collection is *V*.
2. *E* is the set of all connections in the network between the nodes.
3. *P* is a path function that determines the path from one node to another.

4. G is a graph.
5. I_p: set of input functions:

NODE.ID: Displaying the id of the nodes
NODE.CURRENT ROLE: Node acting as a harvester
NODE.DISTANCE: Distance between two nodes initially assigned to be 0
NODE.POSITION: Position of node initially set to 0
NODE.RESERVED ENERGY: Reserved Energy of the nodes initially assigned to 2
NODE.NBHRS: The value of all the neighboring nodes across the harvester node.
O_p: Set of output functions:
x cord: Calculating the value of x-coordinate
y cord: Calculating the value of y-coordinate

Discover neighbors: Finding the next nodes that satisfy the mathematical relationships of distances or measured coordinates.

Suppose a adjacency matrix of a graph G is represented by $A = (Ai \cdot j)$ $n \cdot n$ then:

$$Ai \cdot j = 1: \text{if vertices vi and vj are adjacent}$$
$$0: \text{otherwise}$$

Diagonal matrix shows the degree of the top. If D indicates the graph G matrix diagonally, then:

$$di \cdot j = \deg(vi) \text{ for } i = j$$
$$0 \text{ for } i \neq j$$

Calculation of x and y coordinates using mathematical formulas, and determining the position of adjacent nodes and distance to harvesting nodes. The energy is supplied through the harvesting nodes if the distance is beyond or limited by the next node.

This is a recursive procedure that applies to all backbone nodes.
Energy Harvester

$$\int_0^T Pc(t)dt \leq \int_0^T Ps(t)dt + Bo \quad \forall \ T \in (0, \infty)$$

Where
Ps(t) - Harvested Power from energy source at time t
Pc(t) - Energy being consumed at time t
Formula for finding x and y coordinates:

$$x \text{ cord} = a + (AREA \cdot X - a) \cdot {}^* rand(N, 1) - eps(RR/sqrt(5));$$
$$y \text{ cord} = a + (AREA \cdot Y - a) \cdot {}^* rand(N, 1) - eps(RR/sqrt(5));$$

IoT Domain Applications

Comparison and making the node in the neighboring list:

$dij = caldist(vertices(i, :), DEPLOYED\ NODES(j) \cdot POSITION);$

%Cal Dist b/w nodes
if $d\ ij <= RR$%Whether the node is within range?
a. $NBHRS(j) = j;$%Add the id of node in neighbor list

5.3.3 Flow Diagrams

The data flow diagram for improved POLY algorithms is illustrated in Figure 5.2(a) and 5.2(b).

5.4 IMPLEMENTATIONS: POLY AND IPOLY

The IT algorithm is implemented within the MATLAB setting. The random application of the sensor nodes is 600*600 in the experimental setup. In different network topologies, 50 to 250 nodes are used. Maintenance methods based on energy topology are being used. 25byte is an optimal medium access control layer data packet size. No loss of the packet. EECDS, CDS Rule K, A3 and POLY are

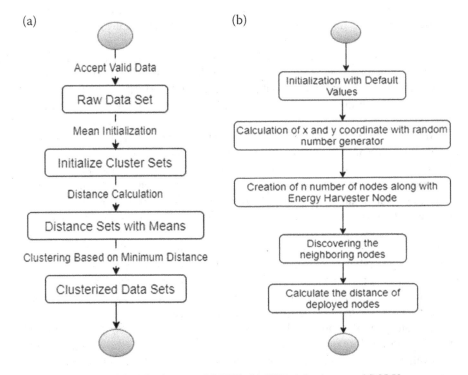

FIGURE 5.2 (a) DFD 0 for improved POLY. (b) DFD 1 for improved POLY.

compared with IPOLY for message and energy overhead. Looking at the stars and mathematical equations it shows that the A3 protocol has a profound message and energy overhead due to the three-way handshake system. IPOLY is more effective than A3 because, during topology construction, it expands less of the energy network. It also has a bigger overhead message than A#. In grid topology, representation nodes are equidistant and have a greater overhead of energy which results in less performance. In comparison to every other protocol based on CDS, IPOLY uses the transmission mechanism to choose the node according to network size.

The following figures illustrate the Polygon Formation for the dense network at steps 1 and 2 shown in Figures 5.3(a) and 5.3(b). Similarly, Polygon Formation for the sparse network at steps 1 and 2 is illustrated in Figures 5.4(a) and 5.4(b).

5.5 FUTURE RESEARCH DIRECTION: IoT RELIABILITY

IoT reliability research was performed at different levels of IoT architecture to improve reliability. This section provides a summary of the available research into system reliability, data quality, network reliability, and anomaly detection, all of which reflect key areas of IoT reliability improvement [7,8].

5.5.1 DEVICE RELIABILITY

Many authors investigating the reliability of IoT devices integrated traditional reliability measures into IoT-centered solutions. Consistency, failure rate, availability, and mean time to repair (MTTR) have been measured. The probable reliability measurement model for connected IoT devices suggests a certain likelihood distribution in the failure structure of IoT devices. Reliability measure $R(t)$ is described by the authors as the probability that the system works perfectly at intervals $[0, t]$. It allows estimating the expected failure time, availability, and reliability of a given IoT system.

MTTR, mean time to failure (MTTF), mean time between failures (MTBF), and availability measures to capture the reliability of heterogeneous IoT devices have been given as a mechanism. This method considered both identified and unidentified types of devices and aimed to distinguish between reliable and unsuitable devices to obtain data from reliable devices and to exclude data from untrustworthy devices. The process included four phases: system identification, classification of specifications, estimation of conviction, and confirmation of conviction. Using this approach, the users were able to establish a ranking of associated fitness devices, based on their reliable outcomes from established consistency methods.

The research presented in this section helps to understand that devices in our IoT organization are reliable and likely to fail. These research pieces help to explain how such knowledge can be quantified using measurements such as availability, MTBF, and MTTR. However, hardware reliability quantification is just one step in the overall reliability outcome. These studies cannot testify to the network stability or evaluate the system's probability of anomalous data or the possibility of spreading the risk.

IoT Domain Applications

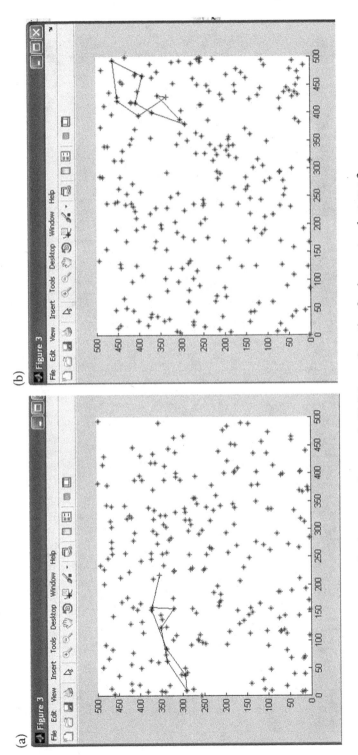

FIGURE 5.3 (a) Polygon formation for dense network at step 1. (b) Polygon formation for dense network at step 2.

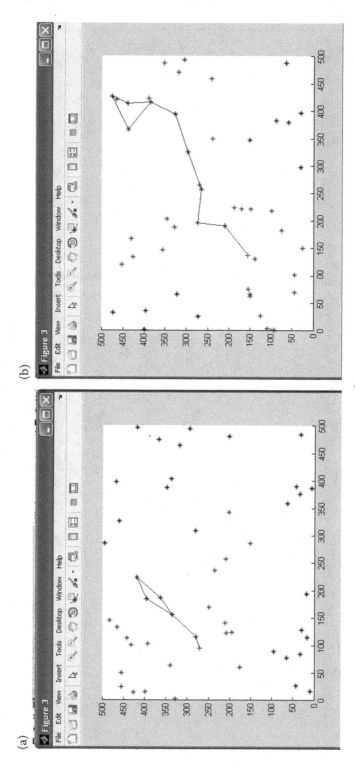

FIGURE 5.4 (a) Polygon formation for the sparse network at step 1. (b) Polygon formation for the sparse network at step 2.

5.5.2 Network Reliability

We will need to be able to demonstrate the efficiency of network infrastructure that forms the backbone of IoT communication in addition to being able to reason about the performance of our IoT devices. Generally speaking, this section discusses two types of network reliability research; network QoS improvement studies and network reliability quantification. This segment shows the latest research on the reliability of IoT networks.

This section discussed a novel IoT network QoS metric, designing a lightweight, energy-efficient routing protocol to improve and measure consistency, especially in case of emergencies. In case of an IoT emergency, an alarm must be responded promptly to. For the assessment of packet loss and track accuracy, the device uses AJIA (adaptive joint protocol based on implicit ACK). The process depends on the protocol's broadcast existence, in which, posts are sent to all neighboring nodes. Therefore, the neighboring nodes will "overhear" the post. Instead of standard ACK messages, this overhearing feature is used to ensure the reliability of the message. The linkage is then measured by the link quality indicator (LQI) metric which uses the packet loss history of the link to establish the consistency of this route. Additional QoS metrics including delay and packet loss are quantified.

A reliability modeling approach has been implemented with the generalized Petri stochastic net (GSPN). This approach theoretized theoretical mathematical models at the edge nodes for statistics on IoT system results. Time consumption, time of response, failure rate, and time of repair were determined. The measurements only speak of the device's output on the edging layer and provide a very narrow view of network accuracy which is not entirely reliable to IoT.

Research obtainable in this section demonstrates that while some efforts have been made to improve IoT network reliability, equally by increasing QoS in the network and by tracking and quantifying network reliability, currently no research methodology integrates system and network reliability into one framework successfully.

5.5.3 System Reliability

Some research was also carried out to assess system-level IoT reliability. These methods have a high degree of reliability which does not capture the individual details, such as which equipment is accountable for failures or which section of the network is accountable for traffic issues.

The usage of a Markov model has been introduced to forecast IoT device trustworthiness requirements. The model from Markov considered that the request could range from the normal condition to total failure in a range of 15 states. The probabilistic nature of the Markov model allows the system to switch from one state to another and to determine the likelihood of failure at some stage. The model only takes into account the states defined in the model design and is unable to react to new circumstances which are not addressed in the model design.

5.5.4 Anomaly Detection

Given the vulnerable status of IoT systems, their restricted devices, and their highly mobile existence, it is important to know the possible presence of anomalous data in the applications of any framework to measure the reliability of an IoT infrastructure. These anomalous data could have serious implications if the framework layer is not diagnosed and used in sensitive circumstances. The latest research on IoT anomaly detection is presented in this section. IoT-specific detection of anomalies is a challenge because solutions have to be lightweight and can handle the varied range of IoT devices.

A smart home anomaly detection approach has been introduced, which incorporates statistical and machine learning techniques in line with system network behavior. During preparation, the functionality is extracted from the network packet data, then standardized and transferred to a clustering algorithm. These clustered labels would then be classified as ensemble methods to decide the outcome of soft-voting. Mechanical fatigue and physical damage were detected by the authors. However, there is a need for more data and performance measurements to assess if the model is working with a large number of devices.

5.6 SUMMARY

This chapter discussed the idea of the connected dominant set (CDS) with different approaches to manage the topology of networks, minimize overheads, and increase the lifespan of networks. Also discussed are clustering mechanisms. The chapter then discussed the network model to build CDS for a graph G of diagonal matrix D. The energy-efficient algorithm based on connected dominating set is addressed in that we had seen the proposed iPOLY algorithm, mathematical model, and data flow diagrams for improved iPOLY. Lastly, it discussed the implementation of POLY and iPOLY, the experiment is carried out in MATLAB.

Exercise
1. What is connected dominating sets? What are the different approaches to building CDS?
2. Differentiate between POLY and iPOLY.
3. Explain iPOLY algorithm with necessary input and output.

REFERENCES

[1] West, D. B. 2001. *Introduction to Graph Theory*. Prentice-Hall, Upper Saddle River.
[2] Liang, O. 2007. *Multipoint Relay and Connected Dominating Set Based Broadcast Algorithms for Wireless Ad Hoc Networks*. Monash University, Melbourne.
[3] Rai, M., Verma, S. and Tapaswi, H. 2009. A power-aware minimum connected dominating set for wireless sensor networks. *Journal of Networks*, 4(6):511–519.
[4] Rai, M., Verma, S. and Tapaswi, S. 2009. A heuristic for minimum connected dominating set with local repair for wireless sensor networks. In: *Eighth International Conference on Networks, ICN 2009*, pp. 106–111.

[5] Wan, P., Alzoubi, K. and Frieder, O. 2004. Distributed construction of connected dominating set in wireless ad hoc networks. *Mobile Networks and Applications*, 9:141–149.
[6] Cai, Z., Wang, C., Cheng, S., Wang, H. and Gao, H. 2014. Wireless Algorithms, Systems, and Applications. *WASA: International Conference on Wireless Algorithms, Systems, and Applications*.
[7] Moore, S. J., Nugent, C. D., Zhang, S. and Cleland, I. 2020. IoT reliability: A review leading to 5 key research directions. *CCF Transactions on Pervasive Computing and Interaction*.
[8] Al-Nabhan, N., Zhang, B., AlRodhaan, M. and Al-Dhelaan, A. 2012. Chapter 62 Two Connected Dominating Set Algorithms for Wireless Sensor Networks. *Springer Science and Business Media LLC*.

6 Cellular Automata-Based Topology Control Algorithms for IoT Domain Applications

6.1 INTRODUCTION

In WSNs the objective is to choose a suitable subset of nodes that might monitor an area at the lowest energy cost and so prolong their network life. The topology algorithms are based on the selection of a suitable subset of sensor nodes in a deterministic or random fashion that must remain active. The cellular automata (CA) utilize cellular and cyclic self-reproduction systems for leading reenactments to assess to play out these calculations and analyze the impact and part of the area determination in the proficient use of created calculations. Distinctive square determination plans are utilized for selecting the district as a part of self-reproduction relying upon the execution. Piece choice is made amid the upgrade of states over a cross-section of cellars. To begin with, the cella itself is considered as is a neighborhood, however, when the condition of a cella is kept up by utilizing the move work the locale is redesigned.

A range of neighborhood schemes has been investigated for research into how the selection of the area in cellular automaton models can influence the efficiency of topology control algorithm simulations in WSNs.

In cellular automation, a neighborhood relationship is specified across a cell grid and identifies each cell's neighborhood during state updates. For instance, a cell neighborhood can be described as the set of cells at two (i.e. two hops) or less distance from the cell. Each cell changes its state with a transition function/regulatory that incorporates all cell states in its vicinity (which usually includes the cell itself).

We analyze the effect on the performances of a topology control technique in a WSN that different neighborhood applications might have: for example, if each cell contains a sensor, the adopted neighborhood type will be able to restrict the number of active sensors employed to space.

The Moore area of a cell comprises a central cell and eight neighboring cells (Figure 6.1). The Von Neumann neighborhood, as depicted in Figure 6.2 has the central cell and the four horizontal cells. The neighborhood Margolus is the fundamental variety of neighborhoods for cellular automatons. The neighborhood divides the grid into four blocks at each stage. Per cell belongs to two different blocks each step depending on whether the step is an odd number or an even number. Figure 6.3 shows the neighborhood of Margolus. The weighted Margolus neighborhood is a

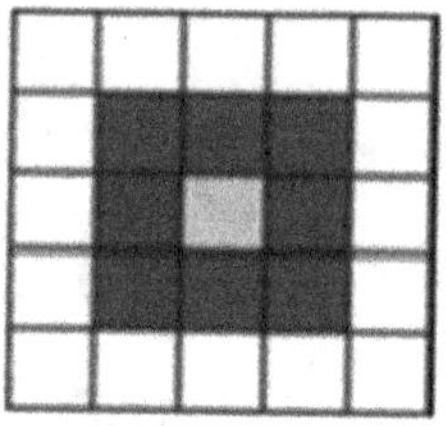

FIGURE 6.1 Moore neighborhood.

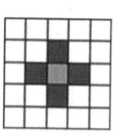

FIGURE 6.2 Von Neumann neighborhood.

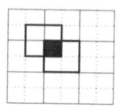

FIGURE 6.3 Margolus neighborhood. Red and blue blocks are applied at odd and even times steps respectively.

variant of the simple, weighty Margolus neighborhood. For each step, the next step is determined by each cell according to not just the neighborhood block it belongs to in the current step, but also the neighborhood block that it belonged to in the prior stage. The neighborhood has a total size of seven cells. The neighborhood of the block is a second variant of the simple neighborhood of Margolus depicted in Figure 6.4. Each of the grid's cells is comprised of four blocks, each of four cells. The algorithm uses every four steps for every block (in the same fashion as Margolus blocks).

The neighborhood weighted block is designed on the main feature of the weighted neighborhood of Margolus. In specifically, we use block's weights. Each cell determines its condition in each step, not just by its current neighborhood block but also according to the neighborhood block that was part of two prior stages. This process also leads to a seven-cell neighborhood. If an area with a slider is supposed to be a slider area, it is divided into three out of three blocks, each having a cell in one block (Figure 6.5). A slider neighborhood uses a combination of a nine-cell neighborhood, which increases awareness about certain cell environments (for example, the Moore region) with the block exchange (like the Margolus neighborhood). The recent study suggests that every node understands its neighbors and their neighbors' positions as wireless network routing algorithms. This concept allows every sensor to recognize the status of their surrounding cells and the status of their neighbors at a distance 2.

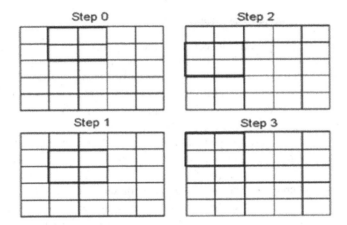

FIGURE 6.4 Block neighborhood.

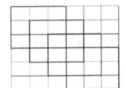

FIGURE 6.5 Slider neighborhood.

6.1.1 CELLULAR AUTOMATA FOR SENSOR NETWORKS

A system of self-reproduction idealizes a physical framework within which space and time are discrete and physical quantities adopt a finite set of values. Lattice will prompt the system of self-reproduction to advance deterministically. Usually, for every cell, the principle of cell condition overhaul is the same and doesn't alter after some time but special circumstances are known. Informally, a self-reproduction system is a lattice of cells, each of which formally is a four-tuple system (C, Σ, N, f), of which C may be in a variety of discrete states (On or Off). The matrix can be measured in any finite number of sets. An area connection is constructed over this grid, showing a d-dimensional cell array or crossing sections of all cells (vectors from ZN list cellars). Σ signifies the letters in order, giving the conceivable expresses every cella may take, N indicates the area (that is, $N \in Z^d$) and f means the move cellars are thought to be its neighbors amid the start-up function of sort $\Sigma N \rightarrow \Sigma$. All cellars are conditioned in time dates. The cella area may be, for example, described as the cell arrangement in which two are separated from the cella or less. Each cell upgrades its condition with a motion to take the conditions of all cells in its vicinity as information at every step. For example, the area of a cella may be determined by the arrangement of two (i.e. two bounces) or fewer cella cells at separation. Each cell updates its state at every step using a move that uses the conditions of the cells in its neighborhood as information.

6.1.2 Sensor Network Clustering

Clustering in the application connection is the appropriate combination of comparable sensors in the collection. The authoritative sensor node structure, i.e. topology, depends on the adjustment of the burden and provides scalability. There are various types of clustering, e.g. close to topology control or change in framework parameters, given the multiple or non-versatile situation. Given the unified and heterogeneous, single or multi-hop type, the clustering is also sortable. The upside of the clustering is that it reduces the number of nodes participating in the transfer when sending complete information to the sink hub. This combination is useful if energy is low, the overhead correspondence for single and multihop systems is scalable and decreased. The great approach of clustering creates spectacular groupings with significant intra-class similarities and little among-class similarities. The nature of the outcomes is determined by the similarities that the approach uses, its use techniques, and its ability to identify the overlapping designs. The partitioning, hierarchical, density-based and model-based approaches are the actual group-based methodologies for applications.

6.2 CELLULAR AUTOMATA-BASED TOPOLOGY CONTROL ALGORITHMS

In WSNs, a random deployment usually covers a region with redundant sensors. When several redundant sensors are active at the same time, the overall network energy decreases quickly and the life of the network is decreased. The primary notion of a WSN topology management technique is that network nodes only have to be active if there are few active, surrounding nodes. Then they persist idle to save their energy.

6.2.1 Cellular Automata Weighted Margoles Neighborhood

The neighborhood of Margolus is the principal range of block areas for self-reproduction. The block divides the cross-section into blocks of four cells at every step. Each cella has two blocks, replacing each time the progression is odd or even, according to each venture. Figure 6.3 displays Margolus's neighborhood. A divergence of standard Margolus quarters that uses weights, is the weighted Margolus neighborhood. Each cella chooses its state at every stage for the next step, not as shown by the area block that it has a position in the present step, as well as the area block it has lived in the past. The block is seven cells in general size. The territory of the block, shown in Figure 6.4, is second to a straight neighborhood of Margolus. Each grid cell is fitted with four blocks, each containing four cells. Each block is used by four steps of the algorithm (in the same design as Margolus blocks).

The neighborhood weighted block was designed to take the notion of the weighted Margolus neighborhood into account. Particularly, we use the weight range block. At each step, each cell decides on its state in the following step,

Cellular Automata-Based Topology Control

according to the area block with a spot in the middle of the current time step as shown by a zone block with two steps in front. The process also leads to a neighborhood of seven cells. The grid is divided into three blocks each three times that share a standard cell when a slider block is approved; these blocks exchange each three-time step (see Figure 6.5). A slider neighborhood connects the top of the 9-cell block that builds knowledge of a particular cella's near environment and the block trade (like occurrences of Margolus neighborhood). In the writing, remote control methods depend on the suspicion that every hub is aware of its neighbors and the positions of its neighbor. According to this hypothesis, every sensor knows the state of its inclusion in separation 2 and the condition of its neighbors.

In WSNs, a zone is normally protected due to its irregular organization by the repeating sensors. In the meantime, when many extra sensors remain dynamic, the system's global energy is dropped swiftly and its system life is shortened. The main idea of the WSN topology control method is that the system node should remain dynamic only when there are few dynamic nodes in the vicinity.

In this study, actualized two essential topology control algorithms must stay active keeping in mind the end goal to broaden the system lifetime, keeping up an ideal scope and network. All the more particularly, nodes choose either to stay active or idle taking into account the dismissal of active nodes in their neighborhood. The self-reproduction system is utilized for the reenactment of TCA-1, and its varieties utilize an n × n grid of cellars. Every cell acid of the grid speaks to a sensor hub and holds data about the sensor position in the framework (dictated through its directions (i,j)), its residual energy to be specific ImpTCA-1 and ImpTCA-2, and must have $Sc_{ij} \in \{0,1\}$ and a clock $Tc_{i,j}$ (represents: counter). Cells may use self-reproduction tentatively considering there be in one of the accompanying two states: 1cellaCij 1 is in 1 state execution. All TCA based ($Sc_{ij} = 1$) when it comparing the system hub holds an active sensor; $C_{i,j}$ on the choice of a fitting subset of sensor nodes incorporates a dynamic sensor, C_{ij} is in state 0 ($Sc_{ij} = 0$) when it is active.

In ImpTCA-1, the choice about the node state (active or idle) is set by the node as per the circumstance of the nodes in their neighborhood, while in ImpTCA-2, this choice is made as far as predefined classifications in which associations have been ordered. The main idea of our topology control methods is that each current sensor node tallies its active neighbors during each progression: if there are intense neighbors, in any event, the node is still idle; the node is still active generally. By l, we signify the most extreme number of sensor nodes that must keep on being active amid every time step expecting a specific neighborhood plan embraced for the self-reproduction system utilized for reenactment. Figure 6.6 shows DFD for ImpTCA1.

6.2.2 Cellular Automata Weighted Moor Neighborhood

ImpTCA-2 intends to choose a suitable subclass of sensor nodes that must stay active to drag out the network lifetime, keeping up most ideal coverage and connectivity.

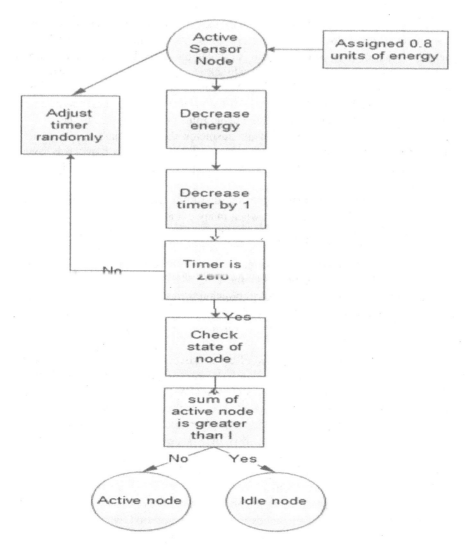

FIGURE 6.6 DFD ImpTCA1.

In any case, ImpTCA-2 utilizes a basic source of randomization keeping in mind the end goal of all the more productively selecting the subset of sensor nodes that will stay active. To replace all nodes with the chance to choose whether they will remain active or idle because of the repeat of active nodes in their neighborhoods, a kind of node group has been chosen and only two-thirds of the nodes of the WSN are hazardously selected to make a choice. The primary inclination behind this trap is that (1) a "small", "settled" group of active nodes is regularly selected, and (2) nodes are encouraged to decide more effectively whether to be active or idle by clearing the excess level surrounding them. The impTCA2 DFD appears in Figure 6.7.

Cellular Automata-Based Topology Control

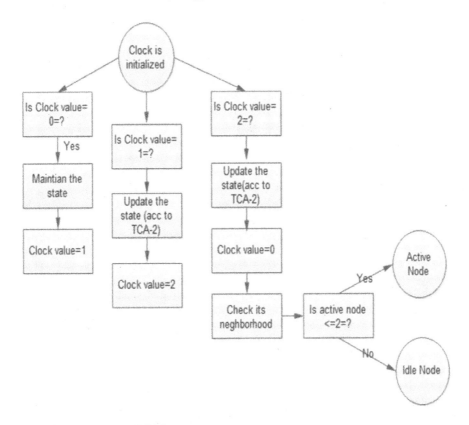

FIGURE 6.7 DFD ImpTCA2.

6.2.3 Cyclic Cellular Automata

Cyclic self-reproduction system (CCA) follows a local rule which is the same for all states S. Each cella in CCA contains different states from state range

$$S = \{0, 1, 2, \ldots\ldots\ldots\ldots k - 1\}$$

Integer k is the extreme number of steps. In the field of Z^2 all cells change their states inside S,

$$I_t(P) = Z^2 \rightarrow \{0, 1, 2, \ldots\ldots\ldots, k - 1\}$$

The CCA generates a spiral structure when cells are varying their states from zero to k−1 as shown in equation (5). The $I_t(P)$ represents the present state of a cella $P \in Z^2$ at integer time t.

$$I_{t+1}(P) = I_t(P) + 1 \mod k$$

144 Energy Optimization Protocol Design for Sensor Networks in IoT Domains

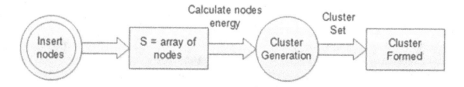

FIGURE 6.8 DFD L0 for ImpCCA.

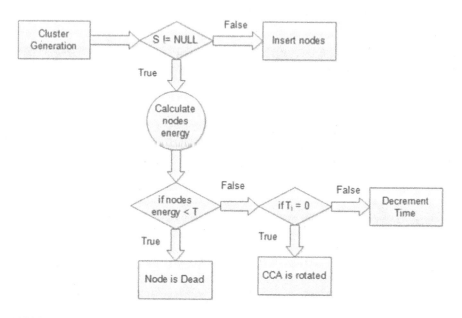

FIGURE 6.9 DFD L1 for ImpCCA.

Cella P changes its state $I_t(P)$ to an alternative state $I_{t+1}(P)$ is shown in equation (6). The DFD of CCA is depicted in Figures 6.8 to Figure 6.10 at different levels.

6.2.3.1 Cellular Automata-Based Topology Control Algorithm

Input:

Node *NODE* {State *S*, Energy *E*},
Initial Energy of each Node *E*,
Energy consumption for each active step E_{ACTV},
Energy consumption for each idle step E_{IDLE},
Number of Active Nodes N_{ACTV},
Number of Ideal Nodes N_{IDLE},
Initial State of Each Node *S*,

//Initialize defaults and initial states for the node
E ç ç 0.8
E_{ACTV} ç ç 0.00006
E_{IDLE} ç ç 0.0165

Cellular Automata-Based Topology Control

N_{ACTV} ç ç 0
N_{IDLE} ç ç 0
S ç ç 1
E ç ç 0.8
PI ç ç 22.0/7.0;
if $NODE.S == 1$ **then**
        $NODE.E$ ç ç $NODE.E - E_{ACTV}$
else if $NODE.S == 0$ **then**
        $NODE.E$ ç ç $NODE.E - E_{IDLE}$

//Active Nodes
if $NODE.S == 1$ **then**

        $\mathbf{N_{ACTV}}$ ç ç ç $N_{ACTV} + 1$

$COVERAGE = N_{\mathbf{ACTV}} * (PI * (RADIO\_RANGE^2))$

//Energy Calculation
R_{BPS} ç ç 1e6// 1MbpS
$DATA\_SIZE_{BYTES}$ ç ç 100// 500 Byes
$DATA\_SIZE_{BITS}$ ç ç ç 100 * 8// 4000 bits
ET ç ç []
ER ç ç 0
$GAIN_{TRANS}$ ç ç 1// Transmitting Gain
$GAIN_{REC}$ ç ç 0// Receiving Gain
$HEIGHT_{TRANS}$ ç ç 1.5// Transmitting Antenna Height
$HEIGHT_{REC}$ ç ç 1.5// Receiving Antenna Height
L ç ç 1// no Loss
LAMBDA ç ç 0.328// Wavelength in Nano meter
E_{ELEC} ç ç ç 50e-9
E_{AMPLR} ç ç 10e-12
E_{BF} ç ç ç 5e-9
$Epsilon_{FRISS\_AMPLR}$ ç ç ç 10e-12// pJ/bit/sq-meter
$Epsilon_{TWO\_RAY\_AMPLR}$ ç ç ç 0.0013e-12// pJ/bit/tetric-meter
PRF ç ç 1e-3
ENR_{DISP} ç ç ç 0
$D_{CROSSOVER}$ ç ç 4 * PI * $HEIGHT_{REC}$ * $HEIGHT_{TRANS}$ * $\sqrt{L}$ / LAMBDA; // 86.2m

//Cluster Head dissipation
D_{HBS} ç ç ç 0 // Distance from head to sender/receiver
ENR_{DISP} ç ç ç $DATA\_SIZE_{BITS}^*(E_{ELEC}^*(N/k)+E_{BF}^*(N/k)+$
$Epsilon_{TWO\_RAY\_AMPLR}^*D_{HBS}^4)$

//Non Cluster dissipation
D_{2CH} ç ç ç 0 // Distance between node and cluster head
$ENR_{DISP} = DATA\_SIZE_{BITS}^*(E_{ELEC}^*+Epsilon_{FRISS\_AMPLR}^*D_{2CH}^2)$;

//Active Nodes
if $NODE.S == 1$ **then**
        $\mathbf{N_{ACTV}}$ ç ç ç $N_{ACTV} + 1$
$COVERAGE = N_{\mathbf{ACTV}}^* (PI^* (RADIO\_RANGE^2))$

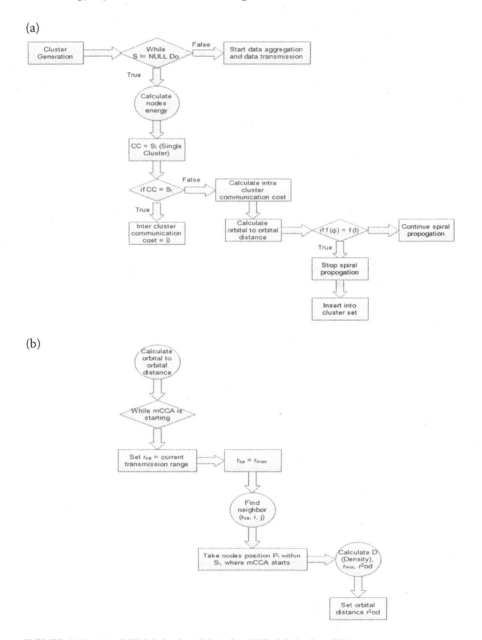

FIGURE 6.10 (a): DFDL2 for ImpCCA. (b): DFD L3 for ImpCCA.

6.2.3.2 Mathematical Model

1. Z^2 = exiting field
2. $I_t(P) = Z^2 \rightarrow \{0, 1, 2, ..., K - 1\}$

   The CCA generate a spiral structure when cells are altering their states from zero to k−1 as equation number 2

Cellular Automata-Based Topology Control

3. $I_t(P)$: Represents the current state of a cell $P \in Z^2$ at integer time t
4. $I_{t+1}(P) = I_t(P) + 1 \mod K$

If and only if for some given threshold value Θ at *Von Neumann* neighbors N(x) a cell P changes its state $I_t(P)$ to another state $I_{t+1}(P)$ at time t+1 is shown in equation 6. The threshold value Θ represents the y number of adjacent cell's conditions in set N(x) within the field $y \in N(x)$.

6.2.3.3 Greenberg-Hastings Model

The Greenberg-Hastings model (GHM) is a basic self-reproduction system that is run in an excitable medium. In GHM according to the state change rule, each cella of the automata changes its state and produced a special type node pattern. The state change rules of the CCA in GHM are

1. If $\gamma_t(x) = n$ then γ_{t+1} (x = n + 1 mode k)

2. If $\gamma_t(x) = 0$ and at least n neighbors are in state 1 then $\gamma_{t+1}(x) = 1$; or else the current state (0) is continued.

$\gamma_t(x)$: Cella's condition (or state) at time t
$\gamma_{t+1}(x)$: Next state of cell at time t+1.

6.2.3.4 Proposed CCA

In the proposed CCA scheme every cella changes its state according to the nine neighbors' cells state conditions. The state change rules of the cells are

1. If cella's present state is $\delta_t(p) = n$ where n > 0 and n < I − 1 then the next state of the cells is $\delta_{t+1}(p) = n + 1$
2. If cella's present state is $\delta_t(p) = n$ where n = I − 1 then the next state of the cells is $\delta_{t+1}(p) = 0$
3. If cella's present state is $\delta_t(p) = 0$ then they check their neighbor cells state and if Θ numbers of nodes are present in the nonzero state then next state is $\delta_{t+1}(p) = 1$ otherwise they are not changing their state $\delta_{t+1}(p) = 0$

where $\delta_t(p)$ is the state of the cells at time t and $\delta_{t+1}(p)$ is the state of cells at t+1 time.

6.2.3.5 Mathematical Analysis

Definition 1: Let X be the monitoring field, covered by the set of sensor nodes, and $f(\emptyset)$ is the intercluster communication cost that is depending on communication distance and node density. Therefore,

1. $f(\emptyset) = \int_{e_i \in M_c} \mu \times \beta_S^C ((\varepsilon + \omega) + \gamma d_i^n) dd_i$

β_S^C: Size of data in cluster c having s number of nodes
μ: Nodes' density

ε: Energy consumed in the transmitter circuit
ω: dissipated energy for data aggregation
γ: dissipated energy in the transmitter op-amp.
e_i: The cluster head collects all data of a single cluster and transmits it to the base station.

Definition 2: The intra-cluster communication cost of the node is $f(I)$ that depends on the cluster size and transmission distance between cluster head and cluster member nodes. Therefore,

2. $f(I) = \int_{e_i \in P_c} \alpha_i(\varepsilon + \gamma d_j^n) dd_j$

α_i: The total bits transmitted along the edge e_i
P_c: Number of nodes in the cluster
d_j: Distance between the transmitter and receiver node in a cluster

Theorem 1: Inter-cluster communication cost $f(\emptyset)$ is single valued and possesses a unique derivation concerning $\emptyset$ all points of a WSN region R is called an analytic or a regular function of $\emptyset$ in that region. A point at which an analytic function ceases to possess a derivation is called a singular point of the function.

Proof:
Let $\omega = f(\emptyset)$ be a single-valued function in a region within the WSN of the variable $\emptyset = x + y$
Then the derivation of $\omega = f(\emptyset)$ is defined the inter-cluster communication cost for data transmission between the cluster head,

$$\frac{d\omega}{d\emptyset} = f'(\emptyset) =_{\partial\emptyset \to 0} \frac{f(\emptyset + \partial\emptyset) - f(\emptyset)}{\partial\emptyset}$$

Theorem 2: Inter-cluster communication cost $f(\emptyset)$ is analytic in the cluster region D between two simple close clusters X and X1, then

$$\int_X f(\emptyset) d\emptyset = \int_{X1} f(\emptyset) d\emptyset$$

where X represents the whole sensor network as a cluster and the X1 is the other cluster within-cluster X.

Proof:
We introduce the cross-cut AB in region X. Then $\int f(\emptyset) d\emptyset = 0$. X1 in clockwise sense and along BA, X in anti-clockwise sense. Therefore, $\int_{AB} f(\emptyset) d\emptyset + \int_{X1} f(\emptyset) d\emptyset + \int_{BA} f(\emptyset) d\emptyset + \int_X f(\emptyset) d\emptyset = 0$ But, since the integration along AB and BA cancel each other, it follows that $\int_X f(\emptyset) d\emptyset + \int_{X1} f(\emptyset) d\emptyset = 0$.

Reversing the direction of the integral around X1 and transposing, we get $\int_X f(\emptyset)d\emptyset = \int_{X1} f(\emptyset)d\emptyset$ here each integration being taken in the anti-clockwise sense. If X1, X2, X3...... be any number of the close cluster within close cluster X then,

$$\int_X f(\emptyset)d\emptyset = \int_{X1} f(\emptyset)d\emptyset + \int_{X2} f(\emptyset)d\emptyset + \int_{X3} f(\emptyset)d\emptyset$$

Theorem 3: Inter-cluster communication cost $f(\emptyset)$ is analytic and if the point a is any node position within X, then energy loss at that node is

$$f(a) = \frac{1}{2\pi} \int_X \frac{f(\emptyset)}{\emptyset - a} d\emptyset$$

$\frac{f(\emptyset)}{(\emptyset - a)}$: Function which is analytic at all nodes position within X except at $\emptyset = a$
a: Center of cluster
R: is the radius of cluster area

Draw a small circle cluster lying entirely within X.
Now $f(\emptyset)/(\emptyset - a)$ being analytic in the region enclosed by X and X1.

$$\int_X \frac{f(\emptyset)}{\emptyset - a} d\emptyset = \int_{X1} \frac{f(\emptyset)}{\emptyset - a} d\emptyset = \int_{X1} \frac{f(a + re^\theta)}{re^\theta} re^\theta d\theta = \int_{X1} f(a + re^\theta) d\theta$$

For any nodes on the network, $\emptyset = a + re^\theta$ and $d\emptyset = re^\theta d\theta$.
In the limiting form, as the circle cluster, X1 shrinks to the node position, as $r \to 0$ we consider every sensor node as a point. The integral approaches to

$$\int_{X1} f(a) d\theta = f(a) \int_0^{2\pi} d\theta = 2\pi f(a) f(a) = \frac{1}{2\pi} \int_X \frac{f(\emptyset)}{\emptyset - a} d\emptyset$$

In general,

$$f^n(a) = \frac{n!}{2\pi} \int_X \frac{f(\emptyset)}{(\emptyset - a)^{n+1}} d\emptyset$$

X is a large monitoring area, and data is traveling among the nodes in the sensor network then $|\emptyset - a| = r$. $\emptyset$ is unevenly distributed within the close cluster X.

$$f^n(a) \vee \leq \frac{Mn!}{r^n}$$

where M is the maximum value of $f(\emptyset)$ on cluster X.

Theorem 4: The analytic function Inter-cluster communication cost $f(\emptyset)$ within the close cluster, the region is average at the center position of a cluster region. This region $f(I)$ is very small. The center position, where inter-cluster communication cost $f(\emptyset)$ is average and intra-cluster communication cost $f(I)$ is small, is the cluster head location.

$$f(\emptyset) + f(I)$$

Proof:
When we consider the cluster of sensor nodes as a circle, then the inter-cluster communication cost of the cluster node $f(\emptyset)$ is average in the center point of the cluster. Because inter-cluster communication cost $f(\emptyset)$ depends mainly on the distance between cluster head and base station. But $f(I)$ also depends on the member nodes to cluster head distance.

$$f(\emptyset) + f(I) \text{ is increased.}$$

We select cluster head as the nearest node of the base station then the inter-cluster communication cost is minimum but within the cluster member nodes and cluster head, distance is increased.

6.2.3.6 Data Aggregation Model
The aggregate of data over a network sensor depends on the correlation of data. We can add a suitable amount of data from nodes via data correlation. Aggregated data are then sent to the head of the cluster (CH). Due to the transfer of information to the base station, a considerable amount of energy is wasted. If the node density has been very high, much energy is spent to convert the same data across a multi-hop sensor network.

6.2.3.7 Entropy Base Model
Model for correlation of data. The entropy-base data correlation and compression algorithm are described in the following equation:

$$B_S(d_0) = b_0 + (s - 1)\left(1 - \frac{1}{\frac{d_0}{C} + 1}\right)b_0$$

d_0: The distance between nodes
b_0: The number of bits from each source
C: Constant parameter characterizing the connection of geographical data

Cellular Automata-Based Topology Control

$B_S(d_0)$: Number of compressed bit messages produced in the s-node cluster from the cluster head. A CCABC entropy-based model is utilized as it reliably and effectively aggregates data.

In CCABC modified equation:

$$B_S(d_0) = b_0 + (s-1) \frac{2e^{\frac{-d_0}{\sigma+s}}}{1-\ln}$$

σ: Minimum size of the cluster.

Other parameters have the same meaning as in the previous equation definition. The correlation error is reduced in our proposed equation. Constraint: − cluster size increases the data correlation error increases. Figure 6.11 shows the rule characterized by David Griffeath. It begins with a uniform random distribution of more than 14 colors, droplets of color waves. The greater part of the underlying flotsam and jetsam are overwhelmed by the droplets. Diamond molded spirals have been made from the cluttered wavefronts. Figure 6.12 and Figure 6.13 show the rules characterized by David Griffeath. Bootstrap alludes to the neighborhood bunches which can proliferate with the assistance of random clamor or different groups. This rule demonstrates a 3-color CCA with contending bootstrap development. Figure 6.14 shows a one-celled critter example to portray the cyclic self-reproduction framework. Figure 6.15, indicates the turbulent phase design for self-reproduction. At the point when the edge is

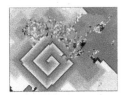

FIGURE 6.11 Basic cyclic self-reproduction system.

FIGURE 6.12 Color bootstrap.

FIGURE 6.13 Cyclic spiral.

FIGURE 6.14 Amoeba.

FIGURE 6.15 Turbulent Phase.

sufficiently high that wavefronts can't twist, yet at the same time sufficiently low that they can progress, CCA rules show a befuddled symmetry stage which consolidates the little scale structure of unsuccessful centers with huge scale cluttered fronts. Our liveliness gives just a look at this conduct, after the run of the mill last length size of a few hundred cells implies that expansive clusters are expected to bolster a feasible relentless state. Little frameworks typically focus.

6.3 FUTURE RESEARCH DIRECTION: CELLULAR AUTOMATA FOR IoT APPLICATION

In response to contemporary issues in cellular automata (CA) modeling, four interconnected theme areas that demand coordinated attention from the CA modeling community have been identified. Four interconnected subject areas are highlighted that deserve concerted attention from the CA modeling community in response to current difficulties in the field. These are: (1) to develop models that capture the multi-dimensional processes of change, such as regeneration, densification and gentrification, in-fill development, shrinkage, and vertical network growth; (2) to develop models that incorporate individual human decision-making behavior into the CA analytic framework; and to develop models that incorporate individual human decision-making behavior into the CA analytic framework; (3) to use emerging big data sources to calibrate and evaluate CA models, as well as to capture the role of humans and their impact on network change dynamics; and (4) to improve theory-based CA models that thoroughly describe network change mechanisms and dynamics.

However, in all studies, it is noticed that the sort of network considered is homogeneous. More challenges will arise if a diverse network is explored. In that situation, the issue to be concerned about is CA design and evolution. If standard routing protocols are employed, the matter can be handled on a case-by-case basis. Similarly, when the WSN is seen in a broader context, the difficulties of scalability, topology reconstruction, energy issues, and overall deployment become a significant challenge. The security issue is a key IoT issue that has not been adequately addressed, and it is an important study topic in IoT network security. The growing number of attacks on IoT networks necessitates more study into improving lightweight ciphers. For lightweight block cipher development, future studies could focus on reducing key size, employing a more frequent dynamic key, decreasing block size, introducing more plain rounds, and constructing simple key schedules.

The quality of service (QoS) in these networks has not yet been fully established. WSN QoS, on the other hand, is very different from typical network QoS. On these networks, QoS cannot be accurately represented. Various criteria are taken into account depending on their application. Total network coverage, optimal network performance of nodes, network life, and energy utilization are some of the criteria for QoS evaluation. Hardware limitations can be imposed on each node by factors like the system's economics, predicted capabilities, node mass, and the realization of ideas in the real world. As a result, future research might focus on how WSNs are implemented and how each node adapts to its new circumstances.

Concerns about the security WSNs are key aspects to consider when designing topologies and maintaining them. Because many nodes in the network are homogeneous, they have the same anti-attack and anti-malware capabilities in the study and analysis of typical WSN security issues. The greater the likelihood of a node being connected, the greater the likelihood of being infected by a malicious program, and the larger the likelihood of being infected. As a result, research based on node protection according to the degree of a node can be undertaken, such that when the node vulnerability function satisfies the power function, all nodes can maintain high availability, making the entire network more stable.

6.4 SUMMARY

This chapter explored the concept of cellular automata-based topology control as well as several techniques. The concept of neighborhood schemes was also addressed. The cellular automata weighted Margoles neighborhood, cellular automata weighted Moor neighborhood, and cyclic cellular automata were all thoroughly discussed. Finally, future research on cellular automata for IoT applications was discussed.

Exercises
1. What is topology control?
2. Explain in detail the functionality of neighborhood schemes.
3. Describe the concept of cellular automata for sensor networks.
4. Explain clustering in sensor networks.
5. List and explain different algorithms used in topology control.
6. Describe the use case of cellular automata for IoT applications.

REFERENCES

[1] Gardner, M. 1970. The fantastic combinations of John Conway's new solitaire game "Life". *Scientific American*, 223:120–123.
[2] Athanassopoulos, S., Kaklamanis, C., Kalfountzos, G. and Papaioannou, E. 26–28 June 2012. Cellular Automata for Topology Control in Wireless Sensor Networks

using Matlab. *Proceedings of the 7th FTRA International Conference on Future Information Technology (FutureTech 12)*, Vancouver, 13–21.

[3] Neumann, J. V. 1966. *The Theory of Self-Reproducing Automata.* (Edited and Completed by A. W. Burks), University of Illinois Press, Urbana, and London.

[4] Wolfram, S. 1986. *Theory and Applications of Cellular Automata*, World Scientific, Singapore City.

[5] Sloot, P., Chen, F. and Boucher, C. 2002. Cellular Automaton Model of Drug Therapy for HIV Infection. *Proceedings of the 5th International Conference on Cellular Automata for Research and Industry (ACRI 02)*, Geneva, 9–11 October 2002, 282–293.

[6] Athanassopoulos, S., Kaklamanis, C., Katsikouli, P. and Papaioannou, E. 25–28 March 2012. Cellular Automata for Topology Control in Wireless Sensor Networks. *Proceedings of the 16th Mediterranean Electrotechnical Conference (Melecon 12)*, Yasmine Hammamet, 212–215.

[7] Yang, L. Z., Fang, W. F., Li, J., Huang, R. and Fan, W. C. 2003. Cellular automata pedestrian movement model considering human behavior. *Chinese Science Bulletin*, 48(16):1695–1699.

[8] Demirel, H. and Cetin, M. 2010. Modeling Urban Dynamics via Cellular Automata, *Proceedings of the 7th AGILE Conference on Geographic Information Science*, Haifa, 15–17 March 2010, 313–323.

7 Performance Optimization in IoT Networks

The approach for monitoring, computing, and confirming the performance in the system to work in efficient manner is known as performance optimization of the system. The performance optimization process is iterative and continuous to achieve satisfactory threshold of expected behaviors of the system. The performance of the system can be optimized using emergent information in domain with appropriate skills in motivation to natural science computing [1].

7.1 IoT NETWORK ISSUES

In IoT domain, there are several issues that can be resolved by applying appropriate techniques to improve the performance of the system.

7.1.1 Fault Tolerance

The fault tolerance system predicts the system issues in advance and provisions the possible solution to it. For example the airplane system has twin engines to handle emergent situation or a spare tyre in a motor vehicle to assist in situation of flat tyre, by replacing it.

The IoT system can be easily damaged or broken and down the entire network. There are possibilities of failure due to error in communication network, protocol conversion, failure of node, or link breakage. The other reasons can be improper functioning of hardware or software which is installed in the end node or devices. The efficient network topology is the best possible solution to tolerate the performance issue by providing alternate links strategies or extra hardware-software backup support for reliable improvement in the system.

The design of IoT network is based on topology design and interlinked layers of control hierarchy. The components of layer are devices, control units, gateways, computing, and storage control systems. The fault tolerance in IoT networks can be achieved by altering topology design for sub-networks in layers and providing alternate fault-tolerance computing within model to find faults level at micro-level with good accuracy. In case of online transactions, the request is resolved through middleware and then diverting request through load balancer for steadily balance traffic to appropriate data servers for accessing database information. In case of failure, the services may handle by standby servers, as shown in Figure 7.1.

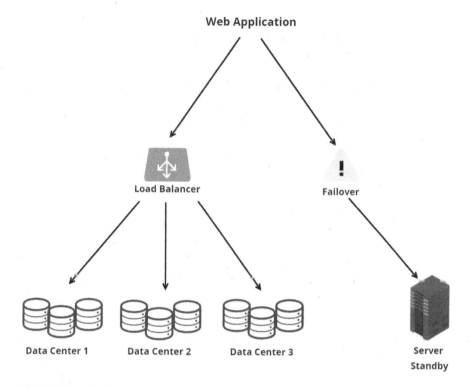

FIGURE 7.1 Fault tolerance system.

7.1.2 Security Enforcement

The security is an essential element in IoT system deployment. The provision of precautionary measures in the connected devices and networks in IoT system is IoT security system. The IoT objects can efficiently work if security at hardware, software, and connectively level is provided effectively. Since IoT involves various computing devices, internet connectivity, and other smart things, need protection system that will resist, detects, and recover from malicious attacks.

The possible action of provisioning security in IoT systems is configuring routers correctly with strong strength passwords and deploying separate Wi-Fi networks, which help the concerned system to work more appropriately. The features mostly not in use can be disabled from time to time, updating hardware and software, setting multiple levels of authorization and authentication levels, and updating firewall policies frequently.

The chances of compromising in IoT-based systems may be possible due to untrusted external cloud support, insecure supportive web, backend application program interface APIs, untrusted mobile interface, or other malicious device connections. Also issues in authorization/authentication, weak encrypting system, and deficient filtering mechanism can lead to an increase in the risk in system.

The attacks in IoT system may also be due to weak security protocols deployed in the system and deficient security policy design. Sometimes security policies are

Performance Optimization in IoT Networks

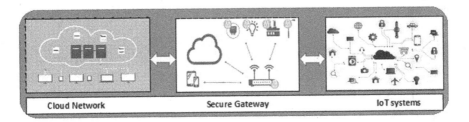

FIGURE 7.2 Security enforcement in IoT enables system.

strong enough but lacking proper documentations lead to inefficient system protections. There are various possibilities for hackers to infect the system by developing malware. The phishing techniques to theft sensitive data, such as corporate workstations and personal devices may lead to face privacy violations due to high profile attacks. The efficient protection mechanisms in cloud storage, gateways, and devices can prevent cyber-attack if concerned manufacturers and security experts access the cyber threat accurately. So every device in IoT system should be alerted while accessing information through gateways from public or private cloud networks, as shown in Figure 7.2.

7.1.3 HANDLING HETEROGENEITY

The integration of various sub-systems in IoT to handle variety of constrained environments is known as heterogeneous IoT system. The system which includes variety of resources like normal computer system, embedded computing nodes, or wireless nodes as a part of wireless sub-network. The heterogeneous network binds various components like terminal units, sensors, and identifying units for acquiring sensed information anytime, anywhere. These networks should connect to any storage units or cloud units through the internet for transferring controls and data reliably.

As the diverse IoT devices are continuously emerging in IoT domain, leveraging heterogeneity is a challenging issue. In heterogeneous networks, the issues are possible connectivity between a variety of connecting devices and networks. The heterogeneous system can be deployed in many areas like smart cities, environmental monitoring, smart homes, smart transportation, etc.

The network structure of IoT is intrinsically heterogeneous, which includes variety of networks like wireless sensor networks, Wi-Fi networks, mobile communication networks, vehicular adhoc networks, etc. In each network type, different communication models get adopted for information acquisition through identifying physical objects or to provide new facility and services. The designing and developments of heterogeneous IoT network system have many challenging issues concern to network architecture and communication technologies. As each IoT base application consists variety of sensors and actuators need to aggregate request-response intelligently through edge devices or gateways to cloud networks for particular services, as shown in Figure 7.3.

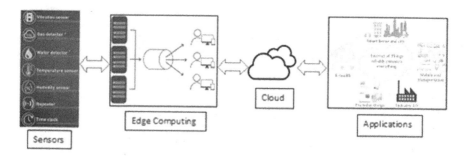

FIGURE 7.3 Heterogeneous IoT network system issues.

7.1.4 SELF-CONFIGURATION

The nodes in IoT networks may have different energy availability, capability, connectivity quality, or priority. The access to physical node is depending as per the context and situation. The challenging issue in IoT networks is adding new node to the system. The self-adoption and self-configuration is complex and limited which possibly not permit due to performance constraints and security issues.

The newly introduced node in the network is automatically configured in self-configuration process by setting in predefining installation procedures as per policies that are guidelines for necessary basic configuration for system operation. The network design manages itself in terms of hardware and software configurations and their resource utilization for self-configuration. The design focus on active monitoring and network node configurations using general communication standards like SSH, SNMP, NetFlow, supported by most hardware manufacturers [2].

The monitoring and execution of tasks for new node to fetch repositories or databases are challenging tasks as shown in Figure 7.4. The other challenging issues in self-configuring IoT networks are bandwidth management which is necessary for quality of services. The emerging multimedia services in IoT networks increased complexity to manage the services in network. The interface issue is another challenge for network initialization in self-configuration process.

7.1.5 UNINTENDED INTERFERENCE

There are various unintended interface aspects such as inter modulator and non-linear effects which lead to other noise-causing issues. The existence of unintended interface issues degrades the performance of IoT networks regardless of licensed or non-licensed frequency band. There is variety of devices that may be Bluetooth enabled, Wi-Fi enabled like phones, laptops, or objects like refrigerators or microwave ovens. These devices operate at different operating standards, but it require to use of spectrum band simultaneously [3].

To operate all IoT devices, as per standard allocated spectrum is shared into multiple bands and each band gets further divided into number of channels. The communication over wireless medium in device communicate one at a time, In case of single system there is no issue, but in case of operating IoT system in many

Performance Optimization in IoT Networks

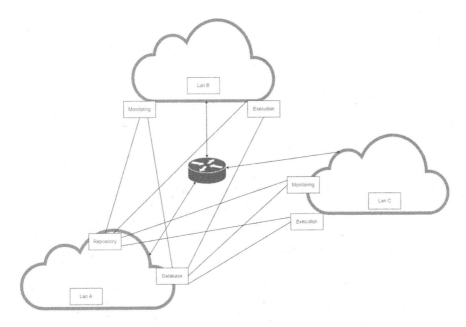

FIGURE 7.4 Self configuration processes.

domains simultaneously, for example, industrial IoT system, its major concern to manage and challenge to maintain performance in the system [3].

The performance in the IoT system may degrade due to unintended interfaces of any device to any other device communication as shown in Figure 7.5. The unintended interfaces may cause slow connections, loss of connectivity in the IoT

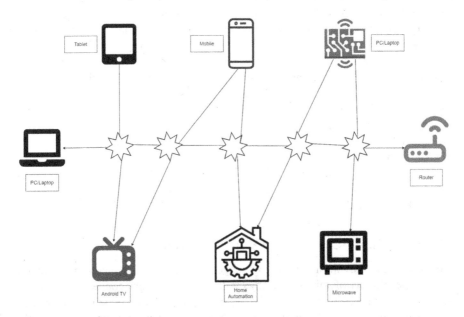

FIGURE 7.5 Unintended interface issues in IoT networks.

network. IoT-enabled devices are growing continuously in emerging systems. The more devices in the IoT networks will cause higher interface issues.

7.1.6 Network Visibility

Since there is an exponential increase in IoT devices in contemporery system like Industrial IoT, the visibility issues increases, also become more complex to resolve the system. To control these diverse number of devices, its necessary to have network visibility for identifying devices in the network. Network visibility helps for monitoring working condition of IoT devices if they are working properly or not or if there are conflicts between access and performances.

There are various visibility tools that give Wi-Fi analytics information for the proper management of the system. The tools also help in identifying and analyzing the flow of data on the radio frequency spectrum which provides information to optimize IoT networks. The network visibility helps the network administrator or manager to visualize the flow of data per application basis and its effect on business-related issues as shown in Figure 7.6. In case of sensitive applications like online banking transactions, the manager can configure the system by setting special policies, setting quality of service threshold at a high level, and optimizing data flow in the network.

7.1.7 Restricted Access

Every Adhoc network deployed is based on primary network for intended purpose. It needs to facilitate or open the access for concerned users only. The IoT device network needs to identify and isolate as per the requirements. The IoT network must be separately accessed from the primary network. The access can be categorized by possibly three types of users, i.e. authorized officials, and dedicated IoT device network administrators and guests access.

The primary network should be used for accessing the sensitive or private data with proper authentication and authorization. Dedicated IoT devices networks should have access to administrators in that domain and guest access for limited generic information access. All these strategies can protect the sensitive data access and protection from attacks.

Further access should be restricted based on source-based firewalls and destination-based firewalls. In source base firewall design, it allows for certain recognized IP addresses, i.e. allowing only users with proper identity. In destination-based firewall design, it allows only certain IP addresses i.e. permitted only certain locations information.

7.2 OPTIMIZATION ISSUES IN IoT NETWORKS

The exponential growth of networking gadgets is tremendously increasing network traffic. The billions of things and objects in IoT system get connected to global networks. Therefore it need to optimize IoT networks for reducing the traffics generated, which indirectly impacts on other services in network for efficiently

Performance Optimization in IoT Networks 161

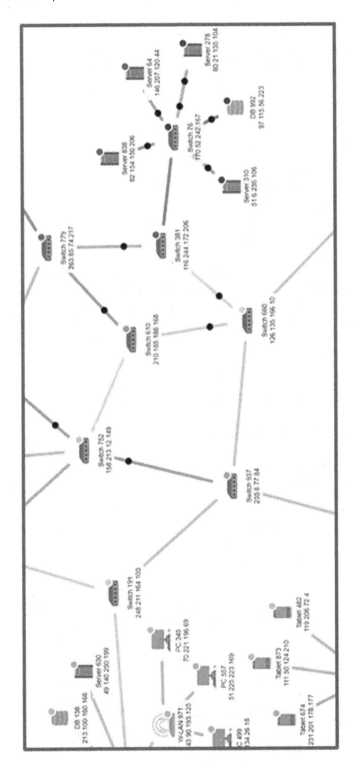

FIGURE 7.6 Network visibility tools.

using network resources. Due to heterogeneous applications and devices in IoT networks, IoT-generated traffic differs from cellular network. It is necessary to regularly monitor traffic generated by IoT networks and its services for optimization purposes. An IoT network consists of integrated devices for variety of services. It needs to control information generated through IoT networks for data exchanges or other transactions, increasing additional burden on network. Hence the efficient mechanism is needed to optimally locate, aggregate data smartly, route data efficiently, within maximum coverage space.

7.2.1 Data Aggregation

As the IoT network consists variety of sensors and other devices, generates heterogeneous data from different sources which required more energy for data transmissions. To reduce energy, the data need to be aggregated, processed, and summarized before sending to destinations. The data aggregations in IoT network have issues due to variety of information like images, audios, and other sensory data. The data modeling and compression can help to optimize traffic in network.

The data in IoT networks are highly reliable due to large numbers of nodes integrated for application in one place. The data reported by various nodes may lead to highly redundant data. So the data send by each node separately consumes huge energy by increasing bandwidth across entire networks, reducing life of network. To resolve the issue data aggregation mechanism is used, which helps to eliminate excessive data transfer, reduce traffic in network, reduce redundant data and increase lifetime of the network. The data aggregation mechanism as shown in Figure 7.7.

There are several applications now ported through IoT platform for society applications like home automation, automatic car parking, controlling HVAC systems, etc., and in communities like monitoring and controlling health issues, updating data for smart cities, improving production efficiencies of factories, etc. All these IoT-based systems collect and aggregate the valuable data and create uniform space for interaction between various IoT devices and their applications. By processing data locally by each device processor, transmit data securely to the uniform space of interaction. The uniform space also receives data from various

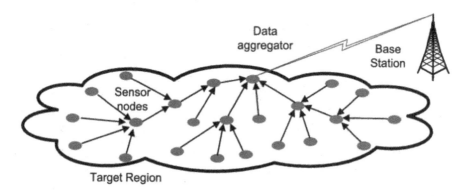

FIGURE 7.7 Data aggregation in IoT networks.

devices and shares it with their intended applications. In advanced applications like vehicular ad-hoc network, where every vehicle connects to IoT for controlling vehicle conditions and security, data aggregation becomes more important. For example on a long-distance run, a driver receives warning message regarding a vehicle that needs urgent attention to repair, but if the driver is not able to rectify the exact message may lead to an accident.

7.2.2 Routings in IoT Networks

In IoT domain routing it is challenging to accomplish energy efficiency where resources are constrained in networks. Also, issues concern to utilizing resources efficiently and designing a routing strategy. The routing operation in IoT network plays a vital role as each node not simply send the data but also forward it, hence routing is the crucial function for competent data delivery over the network. The routing protocols also need to focus on communications reliability with energy efficiency. In most of the IoT-based applications, large numbers of sensors are deployed in terrain environments where battery replacement or recharge is difficult. In routing optimizations, reducing the energy consumption by intelligently handling the routing process for enhancing network life is an important aspect. The maximum energy gets consumed during transmission and reception of data in resource-constrained network since the rest of the time nodes are in either sleeping or inactive mode in network [4,5]. The simple node i.e. end node communicates through internet node or router for a simple message or internet message or control message as shown in Figure 7.8.

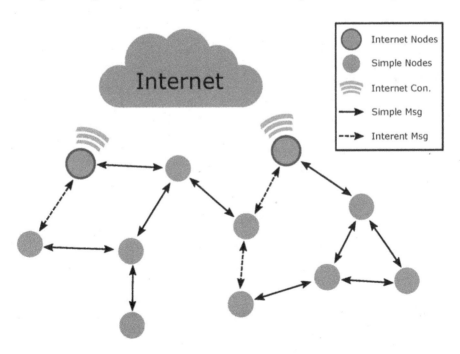

FIGURE 7.8 Routing in IoT networks.

The default routing standard in IoT networks is the routing protocol for low power and lossy network (RPL). The RPL has acquired much more advancement recently but still, there are issues concerns to energy optimizations. In RPL the default objective function for routing decisions is based on only single parameters i.e. link capacity and ignores parameter like energy cost. The routing topology in RPL is based on different links metrics and constraints. The selection of route depends on link quality metric or hop count but most of the time it is unaware of energy resources, hence may select inefficient route. Also, construction of topology for routing based on the single metric makes the network imbalance and creates unrest potential spots near to root node [4,5].

The RPL is a bidirectional communication protocol for communicating between source and sink nodes. Also support for multimode operations such as multipoint to point, multipoint to multipoint, and point to multipoint. RPL interoperates with the IPV6 at the network layer and constrained application protocol (CoAP) at the application layer. The selection of the best route in RPL is based on specific metric parameter like either reliability or delay; hence the selected route inevitably suppresses other parameters effect. For example, if delay metric is considered, the energy consumption does not get recognized. Hence it needs to consider multiple parameters as an objective function for computing efficient path decision. There are various versions of RPL for energy optimization strategies using duty cycle aspect, QoS requirements for resource efficiency, network reliability, and lifetime of the network [6].

Some of the strategies include efficient data delivery for emergent monitoring in special conditions, where forwarding node is aware about sleep and wake-up states. Other strategies consider duty cycle, link quality, and energy condition of the data sending node for optimal path exploration [4].

7.2.3 Optimal Coverage

The network coverage in IoT networks is one of the major issues as it is a necessary and essential requirement for communication. The coverage is a target area where set of sensors covers the involved node area or all target nodes. The coverage of the entire area for access and control or all target network nodes with coverage of all sensing devices is known as optimal coverage [7].

Since the shape of sensing area is irregular in nature, it is considered as a crucial parameter for the coverage of all sensors, for example two-dimensional geometrical based sensing shapes as shown in Figure 7.9. The actual shape of the sensing area is not standard shape and is complex due to terrain locations. The example of real-life sensing shape of the application area is shown in Figure 7.9(c), but for simplicity in understanding we represent it in hexagonal shape or circular shape shown in Figure 7.9(a). The intention of showing hexagonal structure is pointing and no overlapping nature. The circular shape is more popular as it presents low complexity, but has limitations in creating coverage hole as shown in Figure 7.9(b). The limitation may be reduced by increasing the radius of circle. By increasing the circle radius arises the issue of overlapping regions which may lead to redundant information and wasteful consumption of the sensors battery.

Performance Optimization in IoT Networks

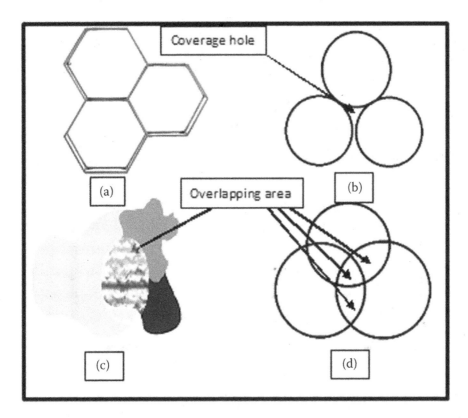

FIGURE 7.9 Coverage area issues.

The reduction of these overlapping sensing regions without coverage holes is the major issue. The area of the overlapping region is directly proportional to redundant information which get sensed by the sensor results into wastages of limited battery of the sensors. The possible solution for minimizing redundancy is to optimize placement of sensor nodes. The optimizing sensor node placement is a single objective function but can be considered multiple objectives for other network parameters to achieve efficient optimizations in the IoT networks.

7.2.4 Sensor Localization

The sensor network is composed of large inexpensive nodes that are densely deployed in a region for measuring certain phenomenon. The process of computing the location of sensor nodes or sensing devices in the network is known as sensor localization. Measuring the location of sensors present in deputed area for special purpose is sensor localization.

There are two phases for computing in sensor localization i.e. distance estimation and position calculation. In the first phase, the relative distance between the base station node and the anchor node is estimated. In the second phase, the coordinates

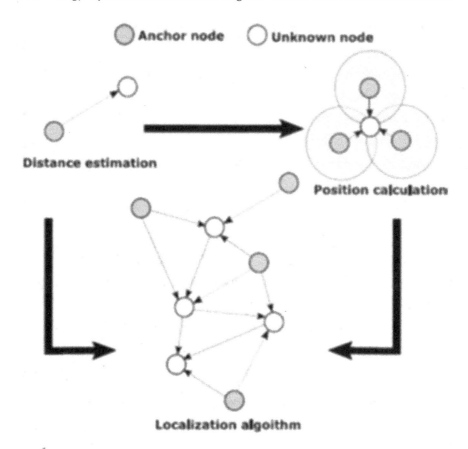

FIGURE 7.10 Sensor localization.

of the anchor node concerning the base node are calculated using gathered information from first phase as shown in Figure 7.10.

In order to track the other nodes position in the sensor network, the distance information and position are manipulated by various localization algorithm strategies. The challenges in sensor localization issues are minimizing the localization errors and increasing the accuracy of the unknown node location [7].

7.3 OPTIMIZATION LEVELS IN IoT

The IoT network structure is classified into three levels as per the data flow in network. The levels are device level, connection level, and management level. Each level can be further classified into sub-levels as per the functions. The device-level controls topology of IoT devices i.e. sensors and actuators in the network, manages middleware work through hardware and software, and also controls local area networks or ad-hoc networks. The connection level is further categorized into controlling distributive network, managing cross-border middleware, and communicating with different machines or devices across the border of local networks.

Performance Optimization in IoT Networks   167

Management level controls data management, manages server-side middleware platforms, and provides necessary services through vertical applications. At each level of system, there are scopes for optimization of energy to prolong the life of the network.

7.3.1 Device Level Optimization

The application of IoT networks as ad-hoc in nature can be deployed for society buildings to monitor data continuously for analysis and automatic controls. It is possible to access aggregate energy consumption from heterogeneous things or objects deployed in the society networks. Most of the time it is difficult to access real-time information data for quick decisions. The real-time data monitoring can help to control energy performance possibly identifying abnormal energy factors to reduce energy consumptions.

The IoT networks use sensors and actuators to sense and control information. We can identify energy consumption data from acquired information through parameters like pressure, temperature, humidity, etc. The administrator monitor and study this parameter information. The data collected helps to understand utilization of the energy as well as implementing preventive and corrective actions. The collected data are also used to evaluate energy consumed by devices, energy cost per device, and parameters affecting major energy consumptions, the structure as shown in Figure 7.11.

The energy consumption data further help to optimize setting of the energy equipments, also the allocation of devices as per the actual need of society. The

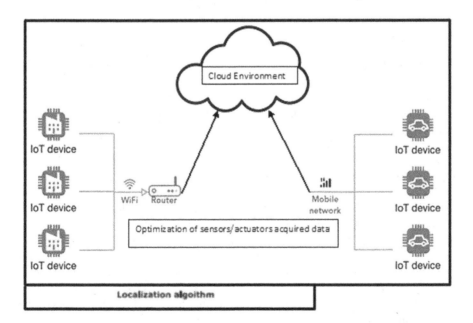

FIGURE 7.11 Device level optimization.

smart meters help to monitor the energy consumption along with alerting optimum usage of energy and relative costs. An automatic heating, ventilation, and cooling system also improve the optimum energy usage during peak and off-hours.

The data collected at device level can further help to analyze at micro-level decisions for necessary action on energy wastages. The efficient energy solutions can possibly not only control simple sensors like temperature adjustors but also more complex control on industrial circuit breaks and building automatic controllers. The IoT energy optimization solution available offers optimum energy management by real-time automatic heating, ventilation, and cooling systems data generation for identification of CO_2 level and temperature which helps further to adjust tuning parameters.

There are various middleware and connectivity protocols like message queuing telemetry transport (MQTT), constraint application protocols (CoAP), and Bluetooth low energy (BLE) protocol for communication in IoT device networks The IoT middleware is responsible for selective data access as per query analysis. The middleware selection depends on the scenario and priority according to demand. The middleware solution available provides qualitative data in the transaction, but it is insufficient for energy consumption. The quantitative data analysis in query transactions can help to compute energy consumption related to power up time in the middleware devices or time required for system computations.

In ad-hoc networks, the energy efficiency of routing protocol affects the performance of the network. There are several parameters that can be contributed to a significant level of optimization for computing network performance. The routing overheads and quality of service metrics with throughputs and delays further impact optimization computing in IoT networks.

7.3.2 Network Level Optimization

The optimization at network level provides unprecedented possibilities to improve optimization model. Since network level is central control of communication, also called as Heart of any network. It helps to understand the overall performance of the network and obtain visibility for energy consumption. Since IoT network is rapidly emerging in almost domains it is challenging for network operators to maintain application working comfortably on the available network infrastructure for intended traditional applications, where emerging IoT devices are sharing the same infrastructure, and need to utilize efficiently as shown in Figure 7.12.

The IoT device networks utilize the same network infrastructure dedicated to traditional data and voice communications, it is challenging for network operators to offer reliable segment traffics, also need to understand the pattern and behavioral analysis. There are various challenges that need to resolve not only for the newly deployed IoT networks but also for their energy consumption. IoT devices works on low energy levels need to reserve separate towers for IoT networks. To analyze power consumption, operator needs to measure IoT device connectivity at micro-level per cell for accessing the capacity requirement and resource reservation considering optimal usage.

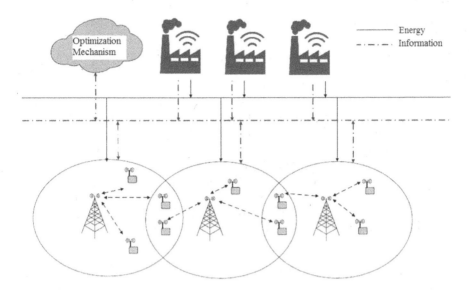

FIGURE 7.12 Network level optimization.

The under usage connections and over usage connections analysis helps to manage optimal reservation of resources and accordingly energy usage. Another issue concern to measuring performance at network level is new device capabilities in a network. The frequent replacement of IoT devices in network also affects planning in network and difficult to analyze connectivity and accordingly measuring energy requirements in the network.

Over a period of time optimization in IoT network is in demand due to exponential growth of traffic from IoT things and objects. Billions of IoT devices are expected to connect to the existing network infrastructure. Hence it is challenging to provide efficient solution by optimizing IoT network to reduce traffic generated by IoT devices, impact on other services in the network, and utilization of network infrastructure efficiently [8].

The traffic generated by IoT devices in networks are different from other regular traffic, as it includes heterogeneous applications and devices. The major difference between IoT network traffic and regular network traffic is that the application in IoT network generates less data traffic but integrates high volume of control message data. Therefore the non-application traffic implies burden on network. Hence it is required to have intelligent techniques for energy optimization by reducing control message communication in IoT networks.

There are several techniques for optimizing IoT network to control message communication. Some of the existing strategies like heuristics methods help with network optimization. Greedy approach is another solution which uses certain assumptions to solve optimization issue. There are various specialized techniques like particle swarm optimization (PSO) algorithm, genetic algorithms, bio-inspired algorithms, evolutionary algorithms, fuzzy logic control algorithm, stochastic algorithms, memetic algorithms, and other miscellaneous algorithms.

The various network parameters used by different techniques are

1. *Prolong network lifetime:* To extend the life of network, load balancing, and failure management of the end node with efficient routing help to prolong the life of IoT network.
2. *Failure management:* Continuous on-off status of node devices in network break links results into signal degradation and reduces network lifetime. Hence there should be minimum linkage failure for reliable communications.
3. *Load balancing:* Balancing the load in a network during routing of data helps to maintain network lifetime. Multipath metric availability makes data communication reliable and less chance to link failures.
4. *Quality of service:* The resource reservations in the network help to provide reliable communication by reducing delays in payload communication and retransmissions.

7.3.3 APPLICATION LEVEL OPTIMIZATION

An application layer in IoT architecture is the highest interacting layer from the clients end. The layer creates the highest interface between network and end devices. The intended applications are deployed at the device end in this layer, for example, browsers that implement application layer protocols like HTTP, SMTP, FTP, etc. For IoT devices, general application layer protocols are not suitable due to extremely heavy weighted and incur large parsing overhead. The specialized protocols for IoT environment networks are MQTT, MQTT-SN, CoAP, XMPP, AMQP, DDS, RESTfull HTTP, LLAP, SOAA, etc. as shown in Figure 7.13. Most of the power consumptions depend on several states in communication. States like CPU, TX-Transmissions, RX-Receiving, LPM-Link power management for linking low power state, most of the power consumed in waking and RX receiving state and list in LPM state.

MQTT – message queuing telemetry transport protocol uses publish-subscribe communication based on TCP/IP protocol and operates in limited bandwidth. It is a

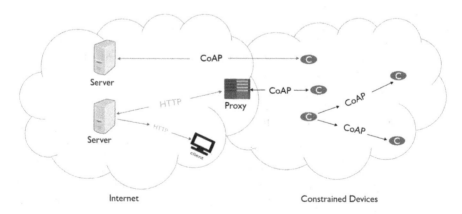

FIGURE 7.13 Application level optimization.

lightweight message protocol designed for sensor networks, which allows device to send or publishes information or data to server. MQTT-SN: MQTT – for sensor networks is a specialized sensor network protocol which is lightweight published/ subscription protocol designed for constrained devices. Like other devices, constrained devices in sensor networks does not send data using TCP/IP stacks to the intender but with MQTT-SN protocol.

MQTT-SN is apparently used in wireless sensor networks as compared to MQTT due to its support for low bandwidth and high link failure, short message length, etc. The protocol is optimized for low cost, limited processing, and storages resources. As compared to MQTT, MQTT-SN reduces the overhead of transmission; since it is ID-based and pre-registered. Only necessary information get exchanged due to message splitting capability. Messages are buffered and later read by clients when device wakeup from offline mode. Due to occasionally initiating and closing transaction process, it optimizes maximum energy [9].

CoAP – constrained application protocol is another protocol specially designed for limited hardware like IoT devices. CoAP is a lightweight protocol which uses limited power for communication end devices and servers. It uses two types of messages in communication i.e. *request* and *response*. These control datagram messages are smaller in size and communicate efficiently among constrained IoT devices without data losses. As compared to MQTT, CoAP is more efficient in terms of power consumption and bandwidth usage while MQTT is more reliable. MQTT has less delay for small packet loss as compared to CoAP, while more delay than CoAP for high packet loss.

XMPP – extensible messaging and presence protocol is another protocol for real-time communication in IoT network and is scalable in nature between IoT devices. As compared to MQTT and CoAP, XMPP consumes more power due to its real-time status for receiving continuous messages. The XMPP protocol is suitable for XML – eXtensible Markup Language-based application which can adjust part header overhead. In case of applications limited to power and bandwidth usage, MQTT is preferable.

RESTfull HTTP – representational state transfer define one or more URL endpoints with a domain, **port**, path, and/or query string HTTP protocol. The protocol is a stateless and interoperable in communication to identify web resources by unique URLs – Uniform Resource Locators. In case of application support HTTP and require REST functionally; the CoAP is most suitable option. Most of the protocols in literature discussed support almost applications in IoT device networks. The properly optimized protocols can effectively used in communication.

7.4 SOLUTIONS FOR IoT NETWORK OPTIMIZATION

The exponential growth of smart gadgets increased relative traffic in network. The number of new IoT devices is getting added daily to the networks for smart applications. An Exabyte of data generated from IoT devices daily in the network increase traffic load on communication system. It is challenging to maintain communication parameters with this amount of data i.e. reliability, routing, quality of services, heterogeneity, congestions, and energy conservation.

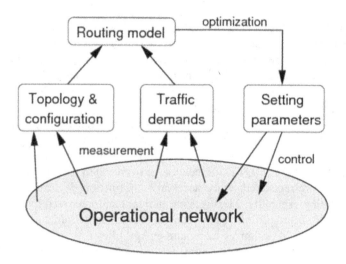

FIGURE 7.14 IoT network optimization parameters.

7.4.1 NETWORK ROUTING

Routing in the network is a process of handling data by choosing selective path across single or multipath network. The data generated through IoT devices should be routed through optimal path toward target destination as shown in Figure 7.14. Routing optimization can be achieved by deploying intelligent routing mechanism. Routing method can be implemented by setting filter to identify duplicate packets and forwarding loops to optimize traffic in network. The method content centric routing (CRC) routes process data aggregation intelligently to reduce traffic in network. The advantage of this method is that it reduces network access cycle by eliminating redundant data. The IoT-based application needs to provide reliable mobile data collection for RPL/6LowPAN protocol with lesser latency packet loss; delay and improved packet delivery rate in the mobile scenario. The other approach for IoT networks i.e. AOMDV dynamically select a stable internet path by regularly refreshing table for internet connection [8].

7.4.2 ENERGY CONSERVATION

To extend the life of the network, energy-saving approach plays important role. The energy optimization mechanism for IoT device network can be achieved by deploying energy-efficient node. For energy conservation design setting the objectives are defined. The optimization engine resolves the objective set to conserve maximum energy in the network. The system is tested through simulation. If required objectives are refined with new target and mechanism gets updated as shown in Figure 7.15.

The optimization engine consists the scheme which continuously monitors the energy level of the nodes to control and enhance network lifetime and throughput. There are sever approaches for energy conservation for network lifetime enhancements. In one of the approach, sending IPV6 packets in BLE capable sensor network

Performance Optimization in IoT Networks 173

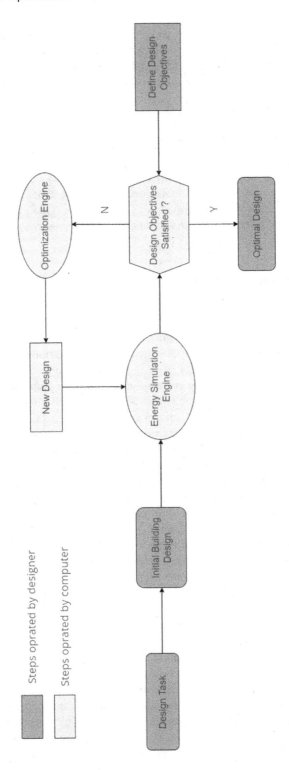

FIGURE 7.15 Energy conservation.

through IPV6 stack implementation by compressing/decompressing header options to conserve energy efficiently transmission in network [8].

In coefficient of variance approach (CoA), the clouds environment and IoT device hardware agree to exchange information about energy and sensors of sleep mode based on battery status. Also predicts maximum information data incoming in the next incoming slots, where additional resources are allocated to avoid delays in the network.

In the approach, self-organizing things (SoT) optimizes energy requirement through self-scheduling algorithms, where least required devices switching to sleep mode and cover the required area with the limited devices. The method works efficiently under traffic load with guaranteed durability. The multitasking environment can be controlled using quality of information (QoA) approach where energy management is used for controlling duty cycle for sensors to achieve quality information and energy management decisions [8].

7.4.3 Congestion Control

Expert predicts 25 billion-plus internet enables devices will get added by 2021 which will potentially raise number of internet connections with traffic loads on the internet. To optimize the traffic on network to conserve energy, efficient congestion control mechanism should be deployed to reduce retransmissions in the network.

The CoCoA – congestion control advance approach removes CoAP restriction on message rate, providing flexible congestion control with protocol security. The CoCoA approach consists of key elements i.e. retransmission time out (RTO), variable back-off factor (VBF), and smaller round trip time (RTT). RTO calculate number of packet required for retransmission which are lost due to network congestion > VBF provide fast retransmission for good connection with small RTT and slower retransmission rate for bad connection with large RTO. RTO monitors these values of transmission and if these values are not updated for longer period then those values get removed from the system [8].

The another approach proposed multilayer solution with parameters like spectrum sharing, data processing architecture, data dimension reduction, and data abandon protocol. Data dimensional reduction is achieved through context awareness and granular computing. The least important data get dropped for reducing congestion in network channel. Another approach uses data offloading system to reduce the congestion in IoT and increase QoS for cellular network (Figure 7.16).

7.4.4 Heterogeneity

As in IoT network, various types of communication mediums are used like Wi-Fi – Bluetooth, Zigbee network, and devices. Applications in network generate heterogeneous complexes and data through the services which take more time to analyze and control for transmission, resulting into consuming more energy. There are several approaches to handle heterogeneous data for energy optimization.

The heterogeneous operations, administration and management (H-OAM) approach proposes failure detection and control mechanism with monitoring and measuring

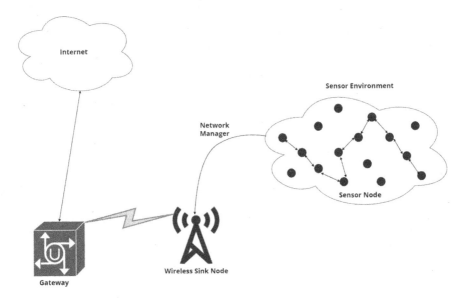

FIGURE 7.16 Congestion control in IoT networks.

performance of heterogeneous network automatically. Since IoT networks consist of heterogeneous connectivity, it is important to monitor and analyze the connectivity continuously. The approach evaluates the variety of data collected from different layers of the communication stack. Another approach SAX – symbolic aggregate approximation uses abstraction framework to optimize sensor data. The mechanism reduces load on the network due to massive data from heterogeneous IoT devices. The approach combines name-based routing for eliminating the IO address assignment problem and helps with content search in large heterogeneous network [8] (Figure 7.17).

7.4.5 Scalability

Due to miniaturization of IoT devices it is possible to deploy IoT networks for almost domains, anywhere in the real-time environment. The enhancements in number of IoT devices generating data traffic in network along with widening network boundaries. So it is challenging to handle this big data to proved efficient network in large scale. One of the solutions to control this data in network is to reduce packet size in the network by compressing packet header. This method is implemented in extensible authentication protocol (EAP) which helps to deploy large-scale devices in the network to support scalability in IoT networks [8].

Another policy is with storage management strategy, to use storage space optimally due to limited storage in IoT device network. The security information is maintained in device node, subset of nodes and after validation, support as ideal storage space. This approach extends scalable trust management for large number of nodes in network resolving scalability issues (Figure 7.18).

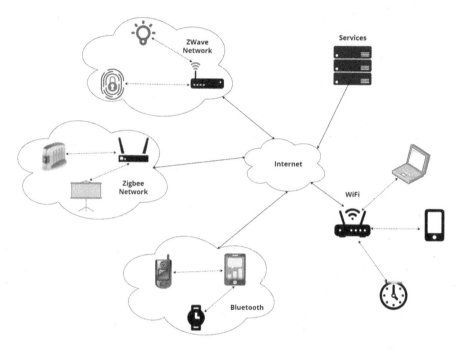

FIGURE 7.17 Heterogeneity in IoT network.

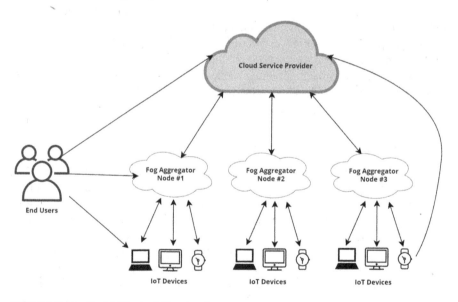

FIGURE 7.18 Scalability in IoT networks.

7.4.6 Network Reliability

IoT network application can be deployed in unmanned area. Reliability management is more important parameter for quality in this type of application. The

Performance Optimization in IoT Networks

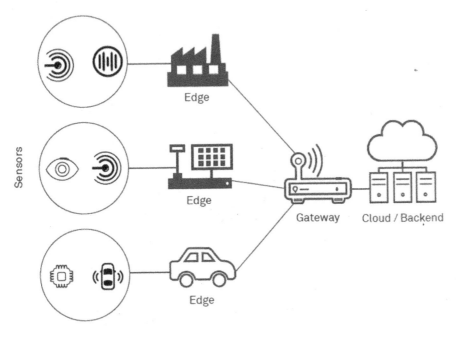

FIGURE 7.19 Reliability in IoT networks.

approach uses lifetime and latency aggregatable metric (*L2AM*) to RPL – routing protocol for low-power and lossy networks for considering minimum path cost during routing in heterogeneous traffic network. The method considers routes based on reliable data and available residual energy. Present in the node [8].

Another approach for decentralized management provides reliable and smarter services for IoT network. The proposed approach uses knowledge and analysis strategies with awaring conditional situations in IoT networks. The approach uses provision confidentiality to protect sensitive data. The RERUM – reliable, resilient, and secure IoT for smart city approach built on network protocol for hardware interfaces. The mechanism increase availability, reliability, security, and trustworthiness in IoT-based networks (Figure 7.19).

7.4.7 Quality of Service

The QoS parameters for IoT networks are bandwidth, packet delays, packet loss rate PLR, avoiding interface, and jitters. It is hard to achieve QoS in sensor networks due to management policy segment and sharing resources in wireless media. DRX – discontinues reception and DTX – discontinues transmission approach for 3GPP LTE to assure energy saving by controlling bit rate, packet delay, and packet loss in network. For optimal resource utilization of LTE Air interfaces, the scheme verifies that after testing the packets of different sizes, packet with smaller size achieves average of the throughput compared to larger size packet. Therefore packet aggregation at the IoT gateway optimizes latency, packet loss, jitter, and bandwidth

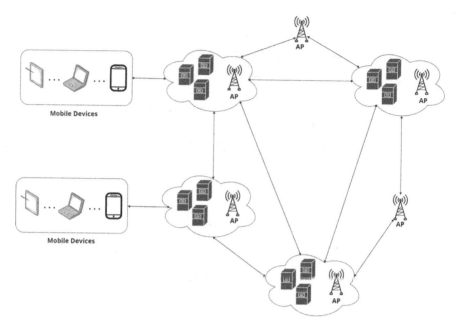

FIGURE 7.20 Quality of service in IoT networks.

utilization for larger small-size packets. Another approach with QoS architecture provides mechanism for controlling transfer and translation from top to bottom layer. Also provide cross-layer management facilities and middleware control for lower layer to provide control mechanism [8] (Figure 7.20).

7.5 SUMMARY

The performance optimization in IoT networks discusses general issues in computer networks for basic understanding such as fault tolerance, security enforcement, handling, heterogeneity, self-configuration, unintended interference, network visibility, and restricted access. The optimization issues in IoT networks and the scope for improvements through various schemes like data aggregation, routing in WSN, optimal coverage, and sensor localization are discussed. Optimizations scope in IoT networks are elaborated through topology level, network level, and gateway levels. Finally, optimal operation parameters in IoT networks such as network routing, energy conservation, congestion control, heterogeneity, scalability, reliability, quality of service, and security are discussed.

Exercise
1. What are general issues in any network?
2. Discuss network issues such as fault tolerance, security enforcement, handling, heterogeneity, self-configuration, unintended interference, network visibility, and restricted access.
3. What are optimization issues in networks?

4. Explain optimization issues in IoT networks: data aggregation, routing in WSN, optimal coverage, and sensor localization.
5. What are various levels of optimization in IoT?
6. Explain optimizations scope in IoT networks through topology, network, and gateway levels.
7. What are different optimal operation parameters in IoT networks?
8. Elaborate on network routing, energy conservation, congestion control, and heterogeneity parameters for optimization.
9. Explore optimization possibilities with scalability, reliability, quality of service, and security.

REFERENCES

[1] Wagh, S. and Prasad, R. 2013. Energy optimization in wireless sensor network through natural science computing: A survey. *Journal of Green Engineering*, 3(4):383–402.
[2] Iqbal, M. U., Ansari, E. A. and Akhtar, S. 2021. Interference mitigation in HetNets to improve the QoS using Q-learning. *IEEE Open Access*, 9:32405–32424, DOI: 10.11 09/ACCESS.2021.3060480
[3] Gupta, A. 2020. Common networking issues with IoT devices and how to avoid them. *Book online IoT 101: An Introduction to the Internet of Thing*. Liverege, USA.
[4] Bhandari, K. S. and Cho, G. I. H. 2020. An energy efficient routing approach for cloud-assisted green industrial IoT networks. *Journals of Sustainability*, 12:1–26. 10.3390/su12187358
[5] Bello, O. and Zeadally, S. 2013. Communication Issues in the Internet of Things (IoT). *Next-Generation Wireless Technologies*, 2013:189–219.
[6] Musaddiq, A., Zikria, Y. B., Zulqarnain and Kim, S. W. 2020. Routing protocol for Low-Power and Lossy Networks for heterogeneous traffic network. *EURASIP Journal on Wireless Communications and Networking*, 2020:1–23. Article number: 21 (2020) Volume 2019 |Article ID 9651915 | 10.1155/2019/9651915
[7] Singh, A., Sharma, S. and Singh, J. February 2021. Nature-inspired algorithms for Wireless Sensor Networks: A comprehensive survey. *Computer Science Review*, 39:100342.
[8] Srinidhi, N. N., Dilip Kumar, S. M. and Venugopal, K. R. February 2019. Network optimizations in the Internet of Things: A review. in *International Journal of Engineering Science and Technology*, 2(1):1–21.
[9] Bernard, M. S., Pei, T. and Nasser, K. 2019. QoS strategies for wireless multimedia sensor networks in the context of IoT at the MAC layer, application layer, and cross-layer algorithms. *Journal of Computer Networks and Communications*, 2019, Article ID 9651915, 33 pages, 2019. 10.1155/2019/9651915

8 Bio-Inspired Computing and IoT Networks

8.1 BIO-INSPIRED APPROACH

Bio-inspired systems use biological phenomena from computing the physical applications. Generally, the bio-inspired system can be implemented using two approaches, i.e. evolutionary system and Swarm intelligence system. The evolutionary system follows evolutionary phenomenon such as reproduction and mutation, while the Swarm intelligence system follows local agent mechanisms to create Ant Colonies or Bee Colonies for global intelligence solutions.

The energy supply in sensor-based networks can be sourced through traditional or renewable systems. There are several energy management approaches for finding optimized energy usage solutions for networks [1]. One of the promising approaches is that they usually avoid searching unpromising local search space to find optimal results. As compared to traditional approaches, bio-inspired approaches are more effective to resolve optimization problems as they effectively search in possible solution space. The bio-inspired process combines imperial observations, experimental, and theoretical studies to synthesize new forms developing new computing paradigm to solve challenging issues, which cannot be solved by the traditional approach efficiently. The process is as shown in Figure 8.1.

8.1.1 Bio-Inspired Computing

Bio-inspired computing is applied for the problem where an efficient and robust solution is required. The approach is used for applications that require optimizations or pattern reorganization computing.

Bio-inspired computing encompasses study domains such as Computer science, Mathematics, and biology as shown in Figure 8.2. The emerging approach to resolve optimization issues is based on inspiration and principles of biological evolution of nature for developing robust computing techniques. The machine learning system uses a bio-inspired approach for creating the optimal solution for complex problems in science and engineering. The non-linear problems with multiple non-linear constraints can be resolved for optimal solutions using machine learning systems.

8.1.2 Bio-Inspired System

The approach which follows analogies from biological creatures in nature for solving engineering problems is bio-inspired systems. Examples of bio-inspired

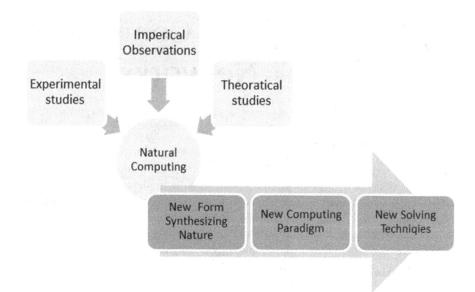

FIGURE 8.1 Bio-inspired process.

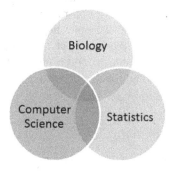

FIGURE 8.2 Bio-inspired computing.

systems are bionic cars, micro robots for mimicking the locomotion of the various insects for crawling, rolling, wall climbing, etc. The major domain of bio-inspired systems solutions is for the systems which are massively distributed and collaborative for example distributed sensing and exploration problems.

The bio-inspired system design works by identifying the suitable biological solution. In the solution, various concern process principles are identified. The concern principles are extracted and patterns of behaviors are observed. The observed pattern then focused on creating an artificial computing model based on behavior strategy. The various principles and artifices are merged to frame solution space. The prototype for the engineering problem solution gets constructed using a bio-inspired system process as shown in Figure 8.2. For the expected and satisfactory outcomes design is tested through prototype based on use cases input concern to problem space (Figure 8.3).

Bio-Inspired Computing and IoT Networks

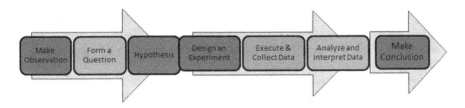

FIGURE 8.3 Bio-inspired system process.

The biological solution can be identified by observations and categorizations of some interesting biological species and solutions of interest. The working of biological solutions for particular biologized problems gets studied and concern principles in the solution steps are recorded. The repletion concerns in the solution space get normalized for framing the unique concern principle solution. The normalized framework gets constructed into an engineering solution for the applicable problem domain. The real application solutions get confirmed by translating and implementing the newly constructed principles as per system requirements.

8.1.3 Bio-Inspired Engineering

The major domain of bio-inspired networking solutions is for the systems where efficient and scalable networking solutions are expected under uncertain conditions for example distributed systems for autonomic organizations. The biological techniques can be effectively used for optimization problem solutions, exploration, and mappings, pattern reorganization, etc. To design a particular solution for the networking domain, one needs to understand the modeling approach. Several technical solutions are mimicking biological simulations. To develop a bio-inspired method in general, have three steps; identifying the analogies which are suitable for problem solving, understanding the logical behavior for modeling, and engineering solution for target problem as shown in Figure 8.4.

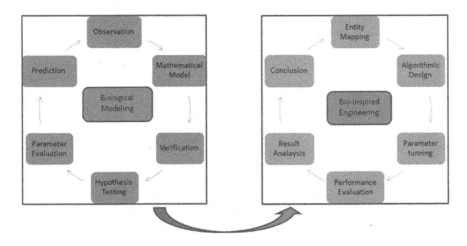

FIGURE 8.4 Bio-inspired engineering.

There are several bio-inspired networking systems based on various networking paradigms. The swarm intelligence and social insects system follow coordination principles in a distributive environment. For direct communication in networks, individual insects' analogies are used, for example, dancing bees. However, in a distributive system, a stigmergic communication analogy is recommended for a continuously changing environment. The individual insects' behaviors for communication follow encoded messages with info-chemical materials rather than radio frequencies. For example, dancing bees' analysis can be applied for routing task allocation and sear4ching peer-to-peer networks in a distributive environment.

The firefly synchronization simulation can be used for precise synchronization in a distributive environment which is a complex issue and hard to achieve in networking. The activator-inhibitor system characteristics are used for analysis or reactor-diffusing mechanism in networking systems. An artificial immune system works effectively for detecting changes in the networking system or identifying deviations from normal behavior in a complex networking system. Information sharing in the distributive network especially in an ad-hoc routing environment is a challenging task. The information dispersion in this contextual for the distribution of information particles may lead to spreading viruses in networks or mobile devices. This can be controlled using epidemic spreading strategies in a bio-inspired system for engineering solutions.

8.2 MOTIVATION FOR BIO-INSPIRED COMPUTING

8.2.1 Self-Organization

The bio-inspired computing approach for developing and managing distributed computing in a dynamically changing environment is a challenging task. The collection of distributed, autonomous and diverse mobile agents designed to mimic the adaptable behavior of the biological system is call self-organization In a self-organized system dynamic modular components are created using mobile agents which automatically migrate and reconfigure while the application is in process of execution. The modular components accumulated can be adjusted as per the local migration scheme of agents using biological behaviors. The self-configuration-based architecture in the bio-inspired system can be achieved through an adaptive architecture system that possibly recognizes and reconfigure agents based on biological processes [2].

The self-organizing architecture allows network protocols to manage the migration of mobile agents in a distributed system. The architecture acts as a middleware in a distributed computing system for adopting self-organization. The self-organization-based architecture enables the components to be exchanged between computing systems allowing applications to be implemented by using these components. The components in this system have their controls and limitations for relative connection between its location and other components' location at connected computing devices [2]. So instead controlled globally, the group of components managed through the association of local components management.

The bio-inspired self-organizing approach can be explained using firefly behavior used for optimization. The firefly approach is a population-based iterative procedure where several fireflies i.e. agents concurrently solve the considered problem of optimization. The communication between agents is performed using bio-luminescent signals which guide an individual agent through the search space. The mechanism of the solution is based on three basic rules of fireflies.

1. Fireflies are unisex, therefore attraction between fireflies is regardless of the sex.
2. The attraction of fireflies id depends on brightness value. The less bright one attracts towards brighter one by minimizing the distance between. If no brighter fireflies are found in the network, less bright fireflies randomly move in the network in search of brighter fireflies.
3. To resolve objective function as optimization, brightness or light intensity of fireflies is resolute.

Each mobile agent i.e. fireflies can either partially involved with other specific parts for a new solution or solely responsible for the execution of skills required to produce a complete solution. Overall in the self-organization process, the evaluation of the process is adopted with network patterns, collective behaviors transformed into system theory, and the final pattern of the algorithm is constructed as shown in Figure 8.5.

8.2.2 Self-Adaptation

In adaption systems, entities can connect or leave the system without disturbing the base system. The adaption systems are self-organized and continue their goals despite adding or leaving users in the system. The entities in adaption systems are

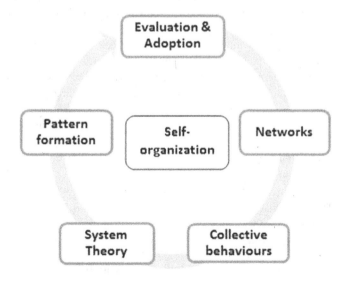

FIGURE 8.5 Self-organization process.

self-evolved and self improves by learning from interaction with the environment. The challenges in adoptions systems are, while the application is in the process allows dynamically best suitable strategies and action as per situation for adaption.

The Characteristics of the network allow to development of existing infrastructure into a more adaptable space including more communication elements, instrumentations, users, and adaptable to different software for analysis and action. There are three phases in adaptive networks; flexible infrastructure, intelligence analysis, and automatic software controls.

8.2.2.1 Flexible Infrastructure

This phase consists of networking physical and virtual elements with automatic transmission and measurement of data from a remote source of wire, radio, or other sources. The flexible infrastructure is a highly intelligent platform that interprets data, so that network decides whether traffic routing intelligently by choosing an optimized path based on appropriate parameters calculations like latency and link capacity.

8.2.2.2 Intelligence Analysis

The flexible infrastructure generates a significant amount of traffic and event data. The data generated uses for learning and adjustment over time in the network. The data used to direct the network for long-term adjustment according to traffic patterns and also alerts which part of the network is vulnerable. The less important data can also be used for analysis to compute the interest of users. The final approval of changes gets confirmed once updated policies get approved.

8.2.2.3 Software Control Automation

Efficient automation of network tasks such as loading access controllers and provisioning routers or automated calculation of configuration of terminal equipment to optimize traffic by resolving congestion [3].

The bio-inspired human immune system, honeybees, ant colonies have several features and organizing principles that can be interpreted for designing and developing adaptive systems. The flexible infrastructure phase can be designed by exploiting the concept of a superorganism that uses self-organizing behaviors and feedback loops for achieving reliability and robustness based on gathered information. The analysis and intelligent phase can be designed by adopting behaviors of honey bees, the waggle dance for computing intelligence using available data to find the optimized route as shown in Figure 8.6. The immune system analogies can help to design a security shield to save from attack. The software control and automation can be implemented by collective combinations of the feedback system [4].

8.2.3 Self-Healing Ability

The biological evolution continues for billions of years based on functional principles in all kingdoms of life for enabling self-healing for diverse types of damages.

FIGURE 8.6 Self-adaption process.

The biological process for self-healing inspired many technical researchers to develop a self-healing or repairing system for problem-solving in the technical domain. The basic prerequisites for converting biological evolution to technical solutions are the involvement of physical, chemical, and mathematical laws that exist in biology and technology.

The observation of natural systems works in humans or plants inspires to study and apply ideology for technical problems solution. The bio-inspired autonomic computing for problem solution manages itself automatically through adaption technologies to resolve system problems and complex maintenance issues in the system. Using a bio-inspired approach for problems in networking can be solved without human intervene by developing an automatic system that can detect and remediate. The approach used for fault tolerance or network-based service management with diverging semantic.

The process of self-healing needs careful investigation. The criticality of a system state can change from normal to abnormal if exceed thresholds, i.e. critical state. In case of exceeding threshold, it is necessary to recover the normal state by starting recovery action. The criticality or deviation can be measured based on desired behaviors, considered functions with specifications, and contexts. If the criticality or deviation from the desired behavior is between the critical state, no healing is required but in case it exceeds beyond thresholds, it's necessary to go systematic healing process.

There are several challenges in self-healing. The challenge like monitoring of relevant events using polling the system state or by a report from the system and receiving events from inside or outside of system temporal or spatial and functional constraints which may cause an intentional malicious attack. In some cases, the

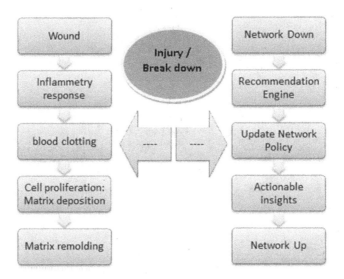

FIGURE 8.7 Bio-inspired self-healing.

system reacts differently in another context. Also, issues with fault analysis and decisions taking for recovery actions.

The natural self-healing analogy applies to self-optimizing networks for real-time network monitoring and identifies faults before they occur in the network. The real-time network data is collected to form the basis of self-healing. The data fed to an algorithm with a machine learning approach, learn from the data sets and identify potential issues and decide the best actions. The action is based on policies set by the network operating server, then model guides to implement actionable insights. The data from events and hosts are fed to the network operating server which controls network operations. The network operating server analyzes based on previous behaviors of the system and improves network management functions by updating policies for changing usage patterns and need of a network (Figure 8.7).

8.3 BIO-INSPIRED COMPUTING APPROACHES FOR OPTIMIZATIONS

The Bio-inspired optimization encompass wide variety of computational approaches which are based on the principles of biological systems. This motivates the application of natural phenomenon to optimization problems. Bio-inspired computing and optimization techniques helps majorly in natural computation. There are many inherited optimization techniques available. This section explains the important bio-inspired optimization techniques based on their development, intention, performance and application as shown in Figure 8.8. Based on bio-inspired strategies, popular bio-inspired optimization algorithms and their application in optimizations is gathered in Table 8.1. The main application in IoT networks concern to optimal coverage, data aggregation, energy efficient clustering and routings, sensor localization are considered.

Bio-Inspired Computing and IoT Networks

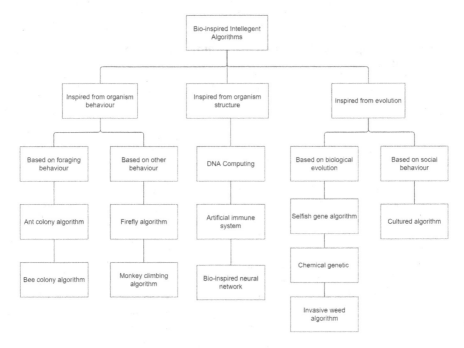

FIGURE 8.8 Bio-inspired intelligent algorithms.

8.3.1 Evolutionary Algorithms (EAs)

The problem which cannot be solved in polynomial time or are an exhaustive and required heuristic approach for solving problems is called an evolutionary algorithm. The combinatorial problems can be easily solved using evolutionary algorithms. The process of the evolutionary algorithm is simple like the process of natural selection. The evolutionary algorithm contains various phases i.e. initializations, evaluation, fitness assignment, selections, and reproductions. The process flow and phases are as shown in Figure 8.9. An approach to finding the best possible solution for the given problem with a particular aspect of natural selections provides implementation in a standard manner. In brief, the better member will outlast and proliferate and the unsuitable member will break down, not contribute further to the gene pool of further propagations [5].

In the initialization phase, an initial population of random individuals is created. The population is the number of possible outcomes or solutions. The elements in the population are called members. It can be either created randomly or based on prior knowledge of the task or a solution that is possibly nearer to expected requirements. The population consists of a wide range of solution space as it needs to represent a gene pool. If we want to explore many outcomes of the process, need a variety of genes in the solution space. In the evaluation phase, the members in the solution space are evaluated as per the best fitness functions. A fitness function is a function that identifies characteristics of members in pools and gives numerical representation in terms of the best possible solution.

TABLE 8.1
Popular bio-inspired optimization algorithms and their application in optimizations

Sr. No	Bio-inspired optimization algorithms	Bio-inspired strategies	Application in IoT networks
1	Genetic algorithm	Evolutionary, Genetic adoption	Optimal coverage, data aggregation, energy efficient clustering and routings, sensor localization
2	Evolutionary programming	Evolutionary, Genetic adoption	Sensor localization
3	Genetic programming	Evolutionary Genetic adoption	Optimal coverage, sensor localization
4	Evolutionary algorithm	Evolutionary, Genetic adoption	Optimal coverage, sensor localization
5	Estimation of distribution algorithm	Evolutionary, Genetic adoption	Optimal coverage, energy efficient clustering and routings, sensor localization
6	Differential evolution algorithm	Evolutionary, Genetic adoption	Optimal coverage, energy efficient clustering and routings, sensor localization
7	Multi-factorial evolutionary algorithm	Evolutionary, Genetic adoption	Data aggregation
8	Ant colony optimization algorithm	Swarm Intelligence, Social behavior	Optimal coverage, data aggregation, energy efficient clustering and routings, sensor localization
9	Particle Swarm Optimization Algorithm	Swarm Intelligence, Social behavior	Optimal coverage, data aggregation, energy efficient clustering and routings, sensor localization
10	Bacterial foraging algorithm	Swarm Intelligence, Social behavior	Optimal coverage, data aggregation, energy efficient clustering and routings, sensor localization
11	Artificial fish swarm optimization	Swarm Intelligence, Social behavior	Optimal coverage, energy efficient clustering and routings, sensor localization
12	Artificial bee colony algorithm	Swarm Intelligence, Social behavior	Optimal coverage, energy efficient clustering and routings, sensor localization
13	Bees algorithm	Swarm Intelligence, Social behavior	Sensor localization
14	Cat swarm algorithm	Swarm Intelligence, Social behavior	Optimal coverage, energy efficient clustering and routings
15	Monkey search algorithm	Social behavior	Energy efficient clustering and routings

TABLE 8.1 (Continued)
Popular bio-inspired optimization algorithms and their application in optimizations

Sr. No	Bio-inspired optimization algorithms	Bio-inspired strategies	Application in IoT networks
16	Firefly algorithm	Swarm Intelligence, Social behavior	Optimal coverage, energy efficient clustering and routings, sensor localization
18	Bee colony optimization	Swarm Intelligence, Social behavior	Data aggregation
19	Cuckoo search algorithm	Swarm Intelligence, Social behavior	Data aggregation, energy efficient clustering and routings, sensor localization
20	Bat algorithm	Swarm Intelligence, Social behavior	Optimal coverage, energy efficient clustering and routings, sensor localization
21	Krill herd algorithm	Swarm Intelligence, Social behavior	Optimal coverage, sensor localization
22	Gray wolf optimizer algorithm	Swarm Intelligence, Social behavior	Optimal coverage, energy efficient clustering and routings, sensor localization
23	Ant lion optimizer algorithm	Swarm Intelligence, Social behavior	Optimal coverage, data aggregation
24	Dragonfly algorithm	Swarm Intelligence, Social behavior	Energy efficient clustering and routings, sensor localization
25	Crow search algorithm	Swarm Intelligence, Social behavior	Energy efficient clustering and routings
26	Lion optimization algorithm	Swarm Intelligence, Social behavior	Data aggregation, energy efficient clustering and routings
27	Whale optimizer algorithm	Swarm Intelligence, Social behavior	Optimal coverage, energy efficient clustering and routings, sensor localization
28	Salp swarm algorithm	Swarm Intelligence, Social behavior	Optimal coverage, sensor localization

The fitness function should be accurate and should give the best outcomes specific to a problem. In this phase, the fitness function is created and this fitness function calculates the fitness of all the members and selects the member with top-scoring priorities.

The fitness function can be single for finding a single optimal point or multiple which can provide multiple optimal points using multiple fitness functions. The set of an optimal solution is called as Pareto frontier which consists of equally optimal elements and no other solution dominates it. In multiple objective functions, the

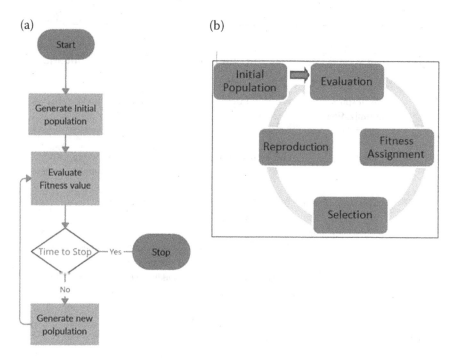

FIGURE 8.9 Evolutionary algorithm: a) Process flow, b) Phases in EA.

multiple objects are targeted to solve the best possible outcomes [5]. In multiple objective functions, once a set of optimal solutions is identified, a decider narrows down the set for a single solution based on the context of the problem of some specialized metric. In reproduction, the phase starts after selecting top members from the earlier phase. These members are used to create the next generation in the algorithm. As per the characteristics of parent members, the children members are created with intermixture qualities of parents. It is tough to create a mixture of quality children members due to the type of data, but using combinatorial problems can mix combinations for valid output. This part of the genetic operator phase is called 'crossover'.

In the later part of the genetic operator phase, new genetic material is added into the generations. This is important because the algorithm may be stuck in local extrema quickly and will not find optimal results. This process is known as 'mutation'. In the mutation part, a small portion of children changed in case they no longer perfectly mirror subsets of the parents' genes. The mutation part is computed probabilistically in which the child receiving mutation and the severity of the mutation controlled by a probability distribution. The algorithm must stop after the validation of a specific condition. There are two ways to end the algorithm, i.e. either the algorithm reached to maximum limits of runtime or an algorithm reached up to the threshold of performance once algorithm end, the final solution is selected and returned [6].

Bio-Inspired Computing and IoT Networks

8.3.2 Artificial Neural Networks (ANNs)

The bio-inspired artificial neural network has several applications like face reorganization, real-time translation, music composition, etc. The concept of the neural network is a base for deep learning i.e. subset of machine learning inspired from the human brain. Neural networks take data input train themselves to recognize the pattern and predict the output. The neural networks are nothing but the computation model of a brain to compute faster than traditional systems. The computations may be related to pattern reorganization, classifications, optimizations, data clustering, etc.

An artificial neural network is a large collection of units that are connected in some patterns and allow communications between the units. The unit represents node or neurons which processes and operates in parallel. Biologically 'neurons' are nerve cells that process information; 'dendrites' are responsible for receiving information from other neurons connected to it, represents input in ANN. The body cell of neurons called 'soma' process information received from dendrites represents the node in ANN. 'Axon' is a communication link through which neurons send information represents output in ANN. 'Synopsis' is the connection between axon and other neurons represents weights or interior connection in ANN. The neurons structure, synapse, and artificial neural network structure is shown in Figure 8.10.

The application of neural networks to IoT devices can help to perform complex sensing and reorganization tasks. The neural networks support processing efficiently and combine sensing data input for a variety of applications in IoT. The energy consumption can be reduced using neural structure optimization techniques and can effectively deploy in resource-constructed IoT devices. The confidence measurement can be computed perfectly for IoT-based networks. Thus neural networks approach supports data collections from various IoT devices, with efficient energy utilization, accurately function with minimal labels. To manage routes in sensor network i.e. IoT devices, neural networks allows the selection of the route that optimize the network performance in a node failure situation. Also, a neural network helps to detect faults in an IoT network with node identifications. The initial network of nodes structure for routing and associated neural network transform structure is as shown in Figure 8.11.

The routing protocol for low power and lossy network (RPL) ensures the reduction of the network energy consumption and maintain network connectivity by adjusting the transmission range of an IoT network. By estimating transmission range with a fixed number of nodes can be computed user prediction model by generating dataset. The node periodically sends a message to the sink node about the amount of energy consumed in the network, by computing the total energy consumed by the network at that time. The efficient transmission range is predicted for a given topology of the network. The IoT application domain likes more cities and smart manufacturing is based on RPL network characteristics composed of resource-constrained devices, which vary in number over time due to node failure or new node insertion into the network.

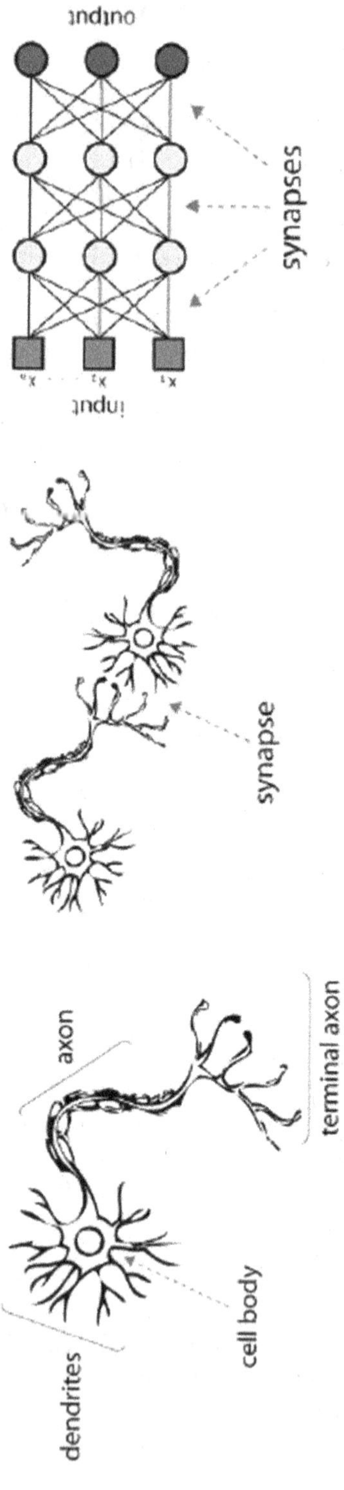

FIGURE 8.10 Neurons structure, synapse, and artificial neural network structure.

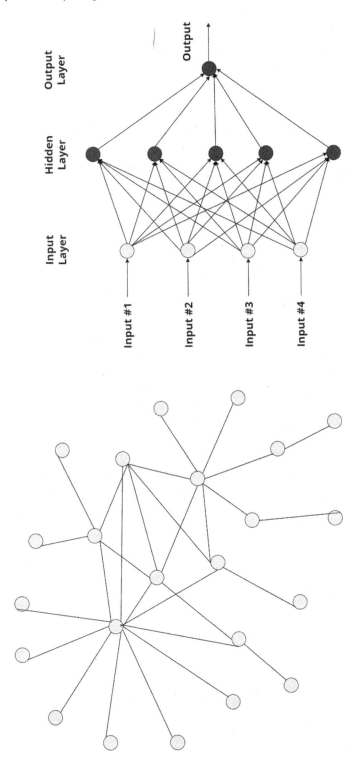

FIGURE 8.11 Artificial Neural Network application for routing.

8.3.3 Swarm Intelligence (SI)

The Swarm intelligence approach is a meta-heuristic method for optimization problem solutions. The technique is inspired by the swarm of birds or 'agents' which interact locally with each other within an environment. The agent behavior in the swarm is simple individually but together they have collective intelligence by following local rules. By following an internal policy of behaviors, the swarm has an emergence of collective intelligence that can solve complex problems intelligently [7]. The recognized swarm intelligence systems are Ant Colony Optimization (ACO), Particle Swarm Optimization (PSO), Artificial Bees Colony (ABC) & Artificial Fish Swarm (AFS).

The Ant Colony Optimization (ACO) algorithm follows the modeling of forgery behaviors of an ant colony. An artificial ant with the artificial pheromone path following the heuristic rule is a statistical random variable that can give an optimized solution by updating pheromone paths. The updated solutions are generated using defined parameters to describe the probabilistic system. The parameters are updated using a previously generated solution so that target promising search space. The ACO is used to find the optimized path in routing networks as shown in Figure 8.12.

Particle swarm Intelligence (PSO) was inspired by the movement of a cluster of birds in search of food. The PSO is a meta-heuristic approach to search solutions from solution space by using a population of candidates or members of the cluster. The population of member solution called particles which move around in the search space or as per particle position and velocity based expression. The particles movement schedule as per the best-known position in the search space, with regular updates for better position information by particles. In the IoT networks, PSOs have well-known application parameters are energy consumption and lifespan longevity for routing purposes [8]. The search space for particles and best particle of the swarm is as shown with phases flow and flow of the process is as shown in Figure 8.13.

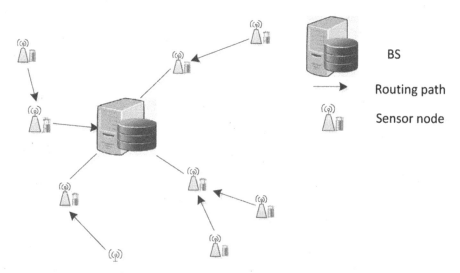

FIGURE 8.12 Ant colony optimization for routing.

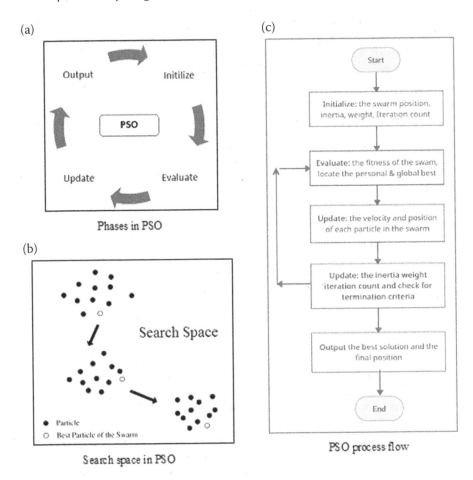

FIGURE 8.13 Swarm intelligence – particle swarm intelligence.

The artificial Bee Colony (ABC) approach represents the intelligent foraging behavior of honey bees swarm to give optimize the solution for the problem. In this model, the colony consists of three groups of bees i.e. employed bees, onlooker bees, scout bees. As per the strategy, there is only one bee artificially employed for each food source[9]. The employed bee goes to their food sources and returns to the hive and dances in the area. The employed bee whose food source is not approved becomes a scout and starts searching for new food space. The onlooker bee watches the dance of the employed bee and chooses a food source depending upon the dances. The scout bee then searches for a new source of food in the solution space [10]. The swarm intelligence approach applied to many optimization problems in sensor networks such as scheduling, cluster analysis, assignment problems, the multi-object optimization problem in sensor networks, etc. Artificial Bee Colony Process and Algorithm is as shown in Figure 8.14.

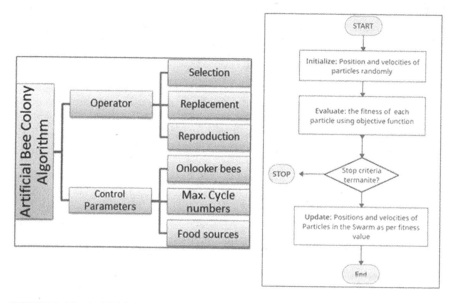

FIGURE 8.14 Artificial bee colony process and algorithm.

8.3.4 FIREFLY ALGORITHM (FA)

The algorithm is inspired by firefly-based or flashing patterns and behaviors of fireflies. The characteristic of fireflies is; since fireflies are unisexual, they are attracted to other fireflies irrespective of their sex. The brightness of fireflies is directly proportional to the attractions. As the distance increases, attraction and brightness decrease. The less brightness will move towards the more brightness one and if nor brighter firefly found, and then particular firefly moves randomly in search of brighter firefly. The brightness of a firefly is intent by the space of the objective function. In the algorithmic process, the relative distances and attractiveness between each couple of fireflies in the population are computed. The position of fireflies is updated as per the latest distance calculated. The new solution and update lighted intensity is determined by the objective function. The fireflies current best is ranked and again computed with updated distance strategy till position and velocities of a particle in a swarm as per fitness value. The scenario is as shown in Figure 8.15.

There are several applications where firefly techniques are used generally in optimization issues. The efficient cluster optimization for optimal path search in routing is computed by variant research techniques in the FA domain. In addition, a chaos-based firefly algorithm for optimization of cyclically large size braced steel drum is used. The modified FA has been used for tracking oil pills and estimating the area. Also, FA is used to discover the opinion leaders in online social networks. The optimization concerning to planning system using computer-based design gives promising solutions. For imbalanced distribution on the network, network configuration is carried using FA whereas optimal switching device placement can be implemented successfully using FA for a power distribution system. The FA

Bio-Inspired Computing and IoT Networks

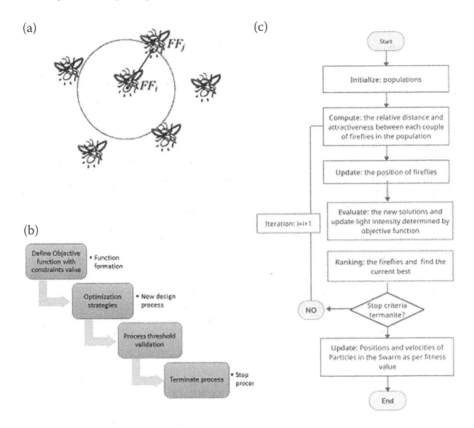

FIGURE 8.15 Fireflies: a) swarm, b) process, and c) algorithm.

algorithm is also used in image processing i.e. for color image segmentation with a multimodal threshold. The modified FA can also be used in Intrusion Detection System for feature selections [11].

8.3.5 Artificial Immune System (AIS)

There are many computational methods and tools inspired by nature termed biologically inspired computing. An immune system is an event-response system occurring naturally which quickly adapts and adjusts to changing situations. The characteristics of an Immune System are alert, identify and neutralize the effect of foreign particles in the body. The biological immune system is the complex network which involves tissues, organs, and cell [12]. The function of an immune system is to recognize foreign elements in our body and respond to eliminate or neutralize them. Several researchers adopt and mimic the strategy for developing new techniques for computational intelligence.

The illness causes due to exposure of the living organism to different microorganisms and viruses. These microorganisms are called 'pathogens'. The body tries to protect against pathogens using different mechanisms, i.e. including high temperature, low pH, and chemicals that resist the foreign elements to grow. The substance that can simulate a specific response of the immune system is called 'antigens'. As the immune system stimulates, it generates several antibodies which respond to the foreign antigens. The immune system distinguishes the original body cells, proteins, and foreign antigens [12]. The process is as shown in Figure 8.16.

The immune system is a multilayer system works for defense mechanism through every layer. There are three main layers, the anatomic barrier, innate immunity, and adaptive immunity. The anatomic barrier layer prevents penetration of pathogens and inhabits most bacterial growth. The pathogen enters into the body by binding and penetrating through the mucous membrane, but this membrane also provides a mechanism of defense with antibacterial and antiviral substances. The innate immunity layer is a composition of physiological barrier, phagocyte barriers, and inflammatory response. The physiologic barrier provides detrimental living conditions for foreign pathogens; for example, low acidity of the gastric system acts as a barrier to infection by ingested microorganisms. Phagocyte barriers kill the antigen and present fragments of the invader's protein to other immune cells and molecules. The inflammatory response allows a large number of circulating immune cells to recruit to the infectious site. Adaptive immunity recognizes and selectively eliminates foreign microorganisms and molecules.

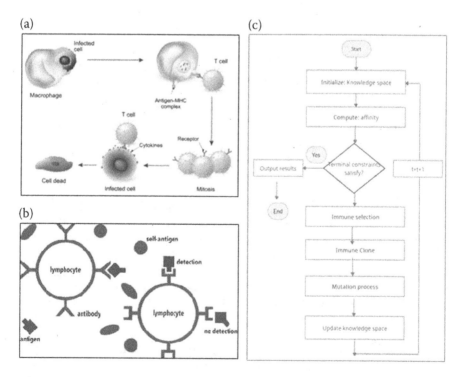

FIGURE 8.16 Artificial immune system: a) living cell system, b) process, and c) flow mechanism.

The computational aspects of the immune system are applied to many engineering applications fields. In pattern matching application the immune system recognizes chemical bindings which depend on the molecular shape. In feature extraction application the immune system recognizes antigen by matching segment of it. The main characteristics of the immune system are to learn through interactive experience based on the occurrence of previous events, secondary response the memory cell that generates a faster and more intense based on previous responses. Hence the mechanism adapted for learning and intelligence enrichments [13].

The immunity system works on distributed processing mechanism as it does not possess a central controller. The detection and response are executed locally and immediately withhold consulting the central organ. Hence this mechanism is easily adopted for distributed computing in a networking environment to control and manage the networks. The immune system works on a self-regulation basis. It responds from low to very strong reaction for protections, while responding to the strong range, use a lot of resources to present the attack. Once the invader is eliminated, the immune system regulates itself by releasing the adopted resources. Hence this mechanism is strongly used in sensor networks for self-organization and configurations. As the immune system has a dedicated protection mechanism, is used in self-protection systems. The artificial immune system-based modeling is being used in many domains like solving optimization problems, computer security, design of intrusion detection system, fault detection and tolerance, pattern reorganizations, distributed learning, sensor networks, job scheduling, recommendations system, etc.

8.3.6 EPIDEMIC SPREADING (ES)

The method which is used to analyze the reason for disease and health outcomes in the population is termed epidemiology. The community and individuals are considered collectively in epidemiology. The branch which deals with transmission and disease control in medical science and maintaining health problems is Epidemiology [12]. The role of Epidemiologists is to study a novel or existing epidemic disease if appears modified, identify design patterns, conduct experimentations, analysis the outcomes as per need, interpret the outcome results and communicate the findings.

In the *design phase*, the epidemiologist identifies the appropriate research strategy and studies design patterns of viruses or bacteria. The justification of the pattern and behaviors are recorded and protocol is designed. The sample sizes are calculated as per the criteria for case selection, appropriate comparison groups are set and questionnaires are designed. In *the conduction phase*, all the appropriate clearance is secured and approvals are obtained conditions to ethical principles. Abstracting records are maintained, the specimens collection and handling are procedures with data management.

In *the analysis phase*, the characteristics of the subjects are described, their progressions are calculated concerning rates and comparative tables are created. The various computation measures like risk ratio or odds ratios, the significance of tests, confidence intervals are recorded. For more advanced analytic techniques the data-driven machine learning, regression analysis, and modeling are used. Finally, in *the interpretation phase*, the findings are presented as perspective, strengths and weaknesses are explored.

The process of epidemic spreading is applied to many network problems i.e. spreading of information in computer networks, propagations of ideas, and rumors in social networks. The epidemic spreading model with effective control strategies applied on a single, multilayer, and temporal networks. The assumptions made in the epidemic spreading model is that a node is capable to interact will nodes in the network at each time step and the infectious probability of a node depends on its connectivity and infectious rate of the epidemic. The computer viruses spread by transmission of data through routers, flu spread through victims who move to different places [14].

The couple spreading process is called the traffic-driven epidemic spreading TDES process. The routing strategies in sensor networks have great influence over this TDES process. The shortest path routing using local routing protocol affects epidemic spreading. The epidemic spreading threshold can be maximized by adjusting the control parameters of the local routing protocol. As compared to the random walk algorithm and greedy routing algorithm strategies, a greedy routing strategy is more effective than a random walk strategy in spreading the epidemic outbreak [14]. The optional value of routing parameters results in a maximum epidemic threshold in the TDES process.

The epidemic spreading model is used to develop effective epidemic control strategies; we can optimize the network structure for suppressing epidemic spreading. In immunization strategy, epidemic spreading is blocked by protecting some critical nodes. In edge removal strategy to suppress the epidemic outbreak of the TDES, properly remove key edges in underlying networks [14]. There are various immunization strategies i.e. the random immunization, degree-based immunization, and betweenness-based immunization gives more strategic immunity as compared to random immunization in sensor networks.

The Module which is popular for epidemic spreading is Susceptible, Exposed, Infectious, Recovered i.e. SEIR is the modified model of SEI which is the simplest epidemiologic model that presents the reality of epidemic growth concerning the populated cases. Since the popularity and comfortable of SEIR paradigmatic model for mathematical epidemiology, the computational model especially to simulate COVID-19 outbreak, SEIR model is used. The advantages of SEIR variant models are that the transparency and accuracy can be achieved and computed through mathematical simulation of biological processes regarding the epidemiological assumptions [15]. By the comparisons of simulated model results and observed patterns, the disease epidemiology is tested.

The model is based on the behaviors of how an epidemic of contagious diseases occurs in the real world [15]. The model is composed of four categories of individuals. S-Susceptible – those who are susceptible to disease, E – Exposed – those who exposed to disease, I – Infectious – those who are infectious and may spread the disease, R – Recovered – those who have recovered from the previous infection and may not spread or catch the disease in future. Epidemic disease spreading and control cycle is shown in Figure 8.17.

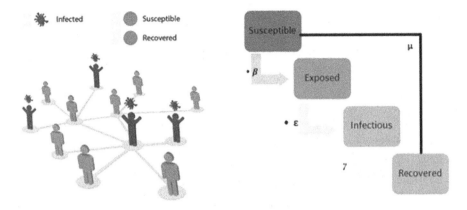

FIGURE 8.17 Epidemic disease spreading and control,

8.4 SUMMARY

The chapter focusses on various approaches in bio-inspired domains like bio-inspired computing, bio-inspired systems, and bio-inspired engineering. The motivation to adopt bio-inspired computing approaches are discussed with characteristics like Self-organization, Self-adaptation, Self-healing ability, Self-optimization, and Self-robustness. The bio-inspired approaches for optimizations are discussed with Evolutionary Algorithms (EAs), Artificial Neural Networks (ANNs), Swarm Intelligence (SI), Artificial Immune System (AIS), Firefly Algorithm (FA), and Epidemic Spreading (ES).

Exercise
1. What are different approaches for bio-inspired domains?
2. Explain the characteristics of bio-inspired computing system.
3. Explain the terms in bio-inspired system
   a. Self-organization,
   b. Self-adaptation,
   c. Self-healing ability,
   d. Self-optimization, and
   e. Self-robustness.
4. What are various approaches for optimization using bio-inspired computing?
5. Explain:
   a. Evolutionary Algorithms (EAs),
   b. Artificial Neural Networks (ANNs),
   c. Swarm Intelligence (SI),
   d. Artificial Immune System (AIS),
   e. Firefly Algorithm (FA), and
   f. Epidemic Spreading (ES).

REFERENCES

[1] Wagh, S. and Prasad, R. 2013. Energy optimization in wireless sensor network through natural science computing: A survey. *Journal of Green Engineering*, 3(4):383–402.

[2] Yeom, K. 2010. Bio-inspired self-organization for supporting dynamic reconfiguration of modular agents. *Mathematical and Computer Modelling*, 52(11–12): 2097–2117, ISSN 0895-7177.

[3] https://www.ciena.in/insights/what-is/What-Is-the-Adaptive-Network-IN.html

[4] Bakhouya, M. and Nemiche, M. and Gaber, J. 2016. An adaptive regulation approach of mobile agent population size in distributed systems. *International Journal of Intelligent Systems*, 31(2):173–188.

[5] Soni, D. Introduction to Evolutionary Algorithms - Optimization by natural selection, towards data science Inc. Canada.

[6] Priya, Annu and Sahana, Sudip Kumar 2020. "Multi-Processor Job Scheduling in High-Performance Computing (HPC) Systems." *FPGA Algorithms and Applications for the Internet of Things*, edited by Preeti Sharma and Rajit Nair, IGI Global, pp. 168–203. doi: 10.4018/978-1-5225-9806-0.ch009

[7] Yang, Xin-She and Karamanoglu, Mehmet 2013. "1 - Swarm Intelligence and Bio-Inspired Computation: An Overview." *Swarm Intelligence and Bio-Inspired Computation*, Editor(s): Xin-She Yang, Zhihua Cui, Renbin Xiao, Amir Hossein Gandomi and Mehmet Karamanoglu, Elsevier, pp. 3–23, ISBN 9780124051638

[8] Azar, Ahmad Taher 2015. "Advanced Metaheuristics-Based Approach for Fuzzy Control Systems Tuning", *Complex System Modelling and Control Through Intelligent Soft Computations* - Springer International Publishing.

[9] Narasimhan, H. 2009. Parallel artificial bee colony (PABC) algorithm. *2009 World Congress on Nature & Biologically Inspired Computing (NaBIC)*, 306–331.

[10] Hong-Mei, L., Zhuo-Fu, W. and Hui-Min, L. 2010 Artificial bee colony algorithm for real estate portfolio optimization based on risk preference coefficient. 2010 International Conference on Management Science & Engineering 17th Annual Conference Proceedings, pp. 1682–1687, doi: 10.1109/ICMSE.2010.5720009

[11] Yang, Xin-She 2021. *Chapter 9 - Firefly Algorithms, Editor(s): Xin-She Yang, Nature-Inspired Optimization Algorithms* (2nd Edition). Academic Press, pp. 123–139, ISBN 9780128219867

[12] Nanda, Satyasai Jagannath, Panda, G. and Majhi, Babita November. 2008. "Improved Identification of Nonlinear Plants using Artificial Immune System based FLANN model", submitted to Engg. Application of Artificial Intelligence, Elsevier, UK.

[13] Chaplin, D. D. 2010. Overview of the immune response. *Journal of Allergy Clinical Immunology*, 125(2 Suppl 2):S3–S23. doi:10.1016/j.jaci.2009.12.980.

[14] Wu, Y., Pu, C., Li, L. and Zhang, G. 2019. Traffic-driven epidemic spreading and its control strategies. *Digital Communications and Networks*, 5(1):56–61, ISSN 2352–8648.

[15] Wagh, S., Wagh, C. and Mahalle, P. 2020. Epidemic computational model using machine learning for COVID-19. *International Journal of Future Generation Communication and Networking*, 13(3):1296–1303 ISSN: 2233-7857 IJFGCN Copyright ©2020 SERSC 1296.

9 Blockchain and IoT Optimization

9.1 BLOCKCHAIN TECHNOLOGY AND IoT

Recently Blockchain technology has attracted significant attention from researchers from academics and industries in various domains. The utilization of Blockchain-enabled IoT solutions in industrial enterprises is increasing in domains like healthcare, energy, real estate, manufacturing, transportations, supply chain management, etc. [1]. Almost all solutions developed by the industry need security enables with strategies. The security with Blockchain mechanism is the promising option for IoT-based application. The deployment of IoT solutions in various domains can be possible with Blockchain for sharing resources and services to intended stakeholders. The possible use cases of Blockchain technology is as shown in Figure 9.1.

9.1.1 Introduction to Blockchain

By using Blockchain, communication for sensitive transactions in business can be automatically cryptographic in line with workflow instead of a manual time-consuming process. Blending IoT design and deployment with Blockchain technology is a security-equipped solution and can prove significant transformations in business.

Blockchain technology has drawn the attention of various industries sectors like health, real estate, finance, government sectors, etc. Blockchain technologies application is operated in decentralized mode instead through a central authority for controlling trustworthiness intermediary. Blockchain enables applications through the network can be deployed safely since stakeholders can interact without worrying about security which boosts the reconciliation between end-users.

The heavily automated workflow in a distributed environment can be possible using self-executing scripts.i.e. smart script contracts in Blockchain. Smart scripts made the Blockchain process highly demanding for researchers and developers working in IoT domains. Although Blockchain usage promising secure solutions still transactions on a decentralized network will always be limited to reasonable or logical.

9.1.2 Blockchain Terminology

The Blockchain can be defined as a chain of information blocks with a digital timestamp to avoid backdate or tamper. The Blockchain is also referred to as Distributed Ledger Technology (DLT) which timestamp any digital record and act transparently through decentralized and cryptographic hashing processes [2]. The Blockchain constructs three important parts i.e. block, node, and minor. The block is

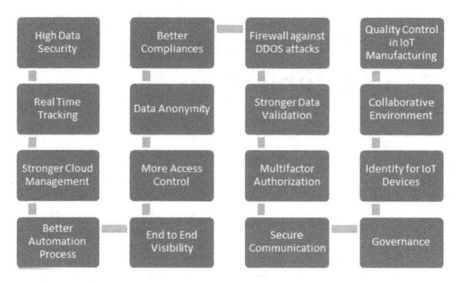

FIGURE 9.1 Blockchain and IoT use cases.

data information with a 32-bit whole number called the nonce and which generates information with block header hash. The hash is a 256-bit number tied to nonce [3].

When the first block of data is created, nonce generates the cryptographic hash. The data in a block is a permanent record tied with nonce and hash unless it is extracted [4]. Minor creates a new block of chain with unique nonce and hash by referring to the hash of the previous block in the chain [5]. As nonce is only 32-bit and hash is 256-bit size, the possible nonce-hash component is roughly four billion possible combinations. Minor uses special software to create nonce, that generates accepted hash. The node can be any electronic device that preserves copies of Blockchain and available for further success with a time stamp record. The node preserves its copy of the Blockchain and approves the newly mined block with updated trust and verification by computing algorithmically. A unique alphanumeric ID is initiated for each new transaction of the document as shown in Figure 9.2

9.1.3 BLOCKCHAIN MECHANISM

Consider a chain of three blocks. Block 1 is the initial block consist no forerunner, block 2 linked to block 1, and block 3 linked to block 2 with their respective hash values [6,7]. Since all blocks are tightly coupled to each other, makes blockchain

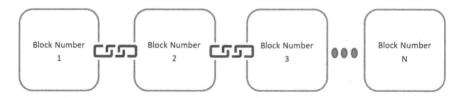

FIGURE 9.2 Structure of blocks in Blockchain.

Blockchain and IoT Optimization

secure. Now if the attacker changes data in block 2, the original hash value will also change and which will not match the hash value linked to block 3. So block 3 will show invalid linkage further chain of block linkage will be invalid.

The hashing technique is an excellent mechanism to avoid tempering the block data. The high-speed computing system today available computes thousands and hash per microseconds and may tamper block with probabilistic comparisons to match hash value as shown in Figure 9.3(a). To avoid this attack, Blockchain uses a new concept of proof-of-work which slows down the creation of the new blocks. The proof-of-work mechanism takes a certain time to configure new values but verification of result is less for confirming final values as shown in Figure 9.3(b).

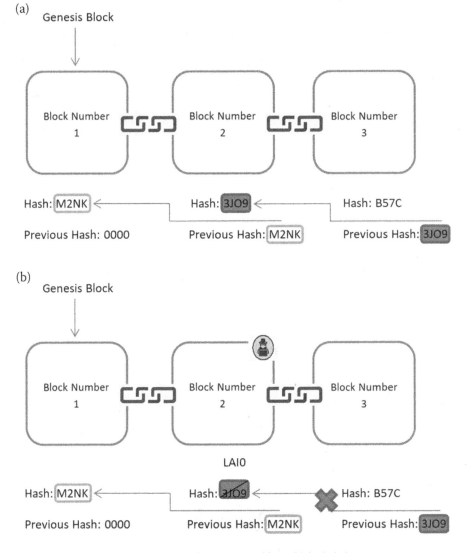

FIGURE 9.3 (a) Working of Blockchain. (b) Working of Blockchain.

For example in bitcoin, it takes almost 10 minutes to compute proof-of-work and add a new block to the chain. If in case a hacker tampers a certain block as in the earlier example he would need at least 10 minutes to perform proof-of-work settings, then only possible to make record change in successor blocks [8].

9.1.4 Distributed P2P Networking

In distributed peer-to-peer networking method, everyone in the network is allowed to join for the transaction. Anybody entering the network gets a full copy of Blockchain. Here instead of using a central authority control system to manage the chain, peers in the network can maintain the chain hence it is more secured.

When a user creates a new block, it gets shared to all nodes in a network. Each user in the network verifies the block is not altered till users complete the verification, then the new block gets added to Blockchain. All the users in the network agree to the opinion of the group for a new block in the chain called consensus. The consensus verifies the validity of the new blocks. The user in the network rejects the block if found tampered or altered. In case of hacker try to tamper Blockchain, a hacker need to make changes to all block in the chain, computing proof-of-work for each block and take control of peer-to-peer network majority, then only tampered block would become acceptable to every node in the network. But it is highly impossible to manage events. Hence Blockchain is in a distributed peer-to-peer network that is highly secured [9].

9.1.4.1 Steps in the Blockchain Transaction
1. Initially someone requests a transaction for a certain purpose may be contracts, records, crypto currency, or other information.
2. With the help of user nodes, the requested transaction is broadcasted to P2P networks.
3. Using the known algorithm, a network of nodes validates the transaction and the user's status.
4. After completion of the transaction the new block is then added to the existing Blockchain, which will be permanent and unalterable.

The steps are as shown in Figure 9.4.

9.1.4.2 Benefits of Using Blockchain
The popularity of Blockchain techniques in almost all sectors is due to the following facts.

FIGURE 9.4 Blockchain transaction.

1. Security: As compared to traditional security mechanism which is centrally controlled for can bring down target system easily. In the distributed ledger system, all users node preserve the copy of the original chain, the system doesn't halt even if some part of network freeze to act.
2. Transparency: As the changes or tampering in the system can view by every peer, it shows great transparency and all the transactions are transparent to all concerns.
3. Decentralized: The information of Blockchain, exchanged by every node as per standard rules. Therefore all record transactions get validated by every node in the network.
4. Collaborative: without mediatory of the third party, allows transaction directly by all user in the network.
5. Reliability: Since the interested participant's verification is confirmed by the Blockchain process, it avoids duplications and makes the transaction process fast.
6. Resilience: In a Blockchain system, the architecture is replicated in a distributive manner, hence it won't get affected majorly in case of attack.
7. Time-efficient: Using Blockchain technique especially in the finance sector, it doesn't need a longer process like traditional for verification, settlement, clearance, etc. An original copy of the agreed shared ledger available to all concerned stakeholders.

9.2 BLOCKCHAIN SUPPORT FOR IoT APPLICATIONS

The emerging of IoT technologies in business increased the business by 13% in 2014 to 30% in 2020. IoT technology is becoming a key tool for business. Some challenges need to resolve for accommodating IoT devices for smart business on large scale. The issues are reliability, latency, congestions, vulnerability transparency, etc. Blockchain technology is steadily maturing since 2017 and now adaptable to work with IoT for security, scalability, and efficient support to its network for associated issues. The major strength of Blockchain mechanism concern to security and transparency, which can resolve issues in IoT enables business that currently facing. Adopting Blockchain mechanism, enhance IoT-enabled business for concerned companies and consumers. The possible IoT applications as shown in Figure 9.5

9.2.1 SECURING IoT NETWORKS

As the IoT devices are exponentially increasing for real-time usage, business related to IoT enable services are increasing tremendously. The various issues associated with IoT solutions are also increasing. For example, smart cities use IoT enables services like connected sensors, lights, smart meters for monitoring and accessing data for analysis. The data acquired may use to improve public utilization, infrastructure monitoring, and other related services. As the IoT devices rely on the network with media like Bluetooth or Wi-Fi, a hacker could attack the network for malicious intent. The attack on the system could impact Manufacturing Maintenance Support. The IoT-enabled system is becoming important in the manufacturing sector.

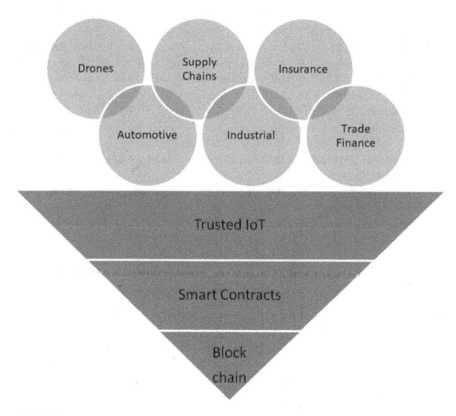

FIGURE 9.5 Blockchain for IoT applications.

The manufacturer uses IoT sensors to check the conditions of machinery in the manufacturing process in a real-time environment system. The IoT-based system enables preventive measures so that they can find and resolve issues before occurrence. The securing components of IoT networks are as shown in Figure 9.6.

The security issues causing connected machinery and IoT devices are more cautious and rolling out the technology in their business. The Blockchain solution in the said domain has perfectly resolved the issue due to its capability to immune in a decentralized structure, secure encryption, and consensus approach. The manufacturer can ensure diagnosed data that is reliable through immutable digital ledger; while sharing a secure record with maintenance partner by saving unplanned halt time associated cost with it.

9.2.2 Manufacturing Maintenance Support

The IoT enabled system is becoming important in manufacturing sector. The manufacturer use IoT sensors to check the conditions of machinery in manufacturing process in real-time environment system. The IoT based system enable preventive measures so that they can find and resolve issues before occurrence as shown in Figure 9.7

Blockchain and IoT Optimization 211

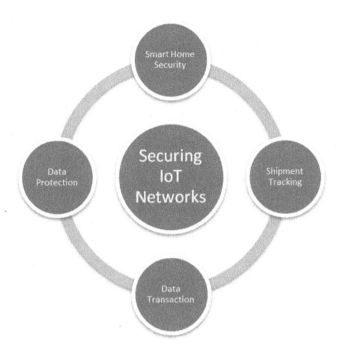

FIGURE 9.6 Securing IoT networks.

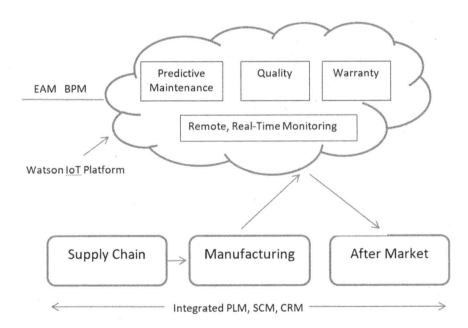

FIGURE 9.7 Manufacturing maintenance support using IoT.

9.2.3 Transparency in Supply Chain

The IoT sensors are used to track and trace goods in a real-time environment in manufacturing and other sectors. The supply chain management maintains an acceptable level of efficiency for monitoring moving parts. The chain continuously grows and challenging to maintain as per expected settings and efficiency.

Blockchain technology addition for supply chain enables information sharing in secure environment accurately and with hack-proof ledger to all stakeholders in the ecosystem. The IoT systems are expanding in almost every domain, including automobiles to aircraft manufacturing and motor-vehicle to smart bicycling. With traditional methods, there are various issues i.e. reliability, slow networks, confidential information, etc. Blockchain mechanism with IoT enables process promise more secure environment for supply chain management system. The Transparency in supply chain is as shown in Figure 9.8.

9.2.4 In-Car Payment Model

The various payment-related services available using the In-car payment system available on display embedded on the car dashboard. The display allows all the activities for automatic transactions instead physical wallet system The In-car payment system allows the driver to pay for fuel, parking, and foods with payment options available on the display unit. The agreement for payment transactions

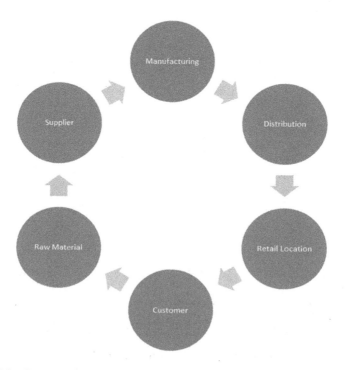

FIGURE 9.8 Transparency in supply chain.

Blockchain and IoT Optimization

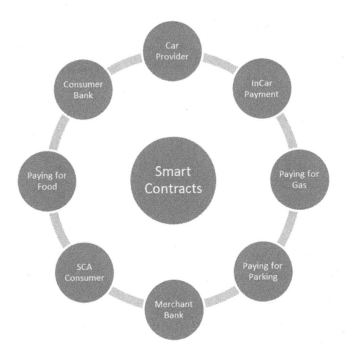

FIGURE 9.9 In-car payment model.

between the In-car display unit and the customer would be registered using a Blockchain mechanism and executed accordingly as shown in Figure 9.9.

The Blockchain mechanism allows the fast transactions of payment with full proof authentication available. Also tracking of transaction history through IoT devices with issuing invoices and payments automatically is possible. The Blockchain confirms interconnection between a variety of platforms, consumers, and target systems, making the process comfortable.

9.2.5 Vehicle Insurance Model

The insurance premium of vehicles gets usually calculated based on average estimates traditionally. The IoT devices connected to the vehicle track the mileage in a real-time environment and may calculate the premium optimally. Binding Blockchain mechanism with this system can be fixed finally also supporting the audit process as shown in Figure 9.10. The Blockchain also assists in tracking drivers' record, vehicle movement records technically. The system can also use to help fraud and theft detections. The possible vehicle insurance model is as shown in Figure 9.10.

9.2.6 Identity Authentication Using Self-Sovereign Identity (SSI)

The digital identity of IoT networks can be preserved using a public key certificate or a cloud-based account. The identity management gets controlled centrally by the IoT management system is as shown in Figure 9.11. The application of Blockchain

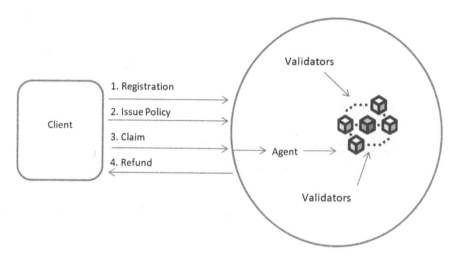

FIGURE 9.10 Vehicle insurance model.

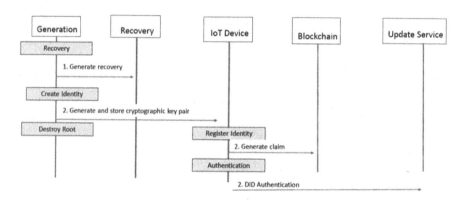

FIGURE 9.11 Managing identity for IoT devices using Blockchain.

for the identification of physical objects, peoples, or organizations is called self-sovereign identity, SSI. In this system, any connected object stores its own identity by ensuring security and privacy in a decentralized network.

For example, the authentication of connected cars for transactions at gas stations can be conformed using credentials like owners' names, vehicle numbers, type of IoT device-embedded, etc. The Blockchain stores credentials of the car to be immutable and can certify by the owner. The SSI system still in maturing phase and advancing its process by adopting more IoT devices lightweight embedded software to the current constrained devices for future applications.

9.3 BLOCKCHAIN WITH IoT NETWORKS CHARACTERISTICS

The data generated through IoT devices can be aggregated decentralized and ensure efficiency using block chain. The rise of IoT devices is get utilized in various IoT

Blockchain and IoT Optimization

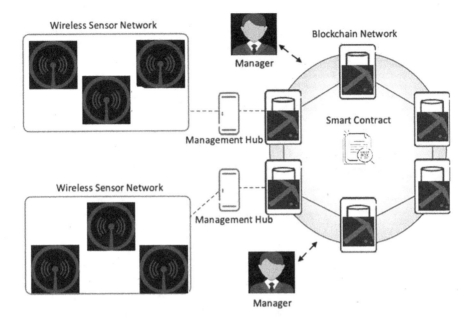

FIGURE 9.12 Blockchain with IoT network characteristics.

domain applications for their simplicity and utility. The combination of blockchain for IoT domains helps to resolve the issues but not complete since both entities are in developing stages. There are various intersections in IoT characteristics and Blockchain technology is as shown in Figure 9.12.

9.3.1 Security

The IoT enables solution providers worried about their data security, information exchanges, and integrity of their physical objects protection from illegal deviation and fictions system. The companies need to always safe and alert from backstairs for securing their products, platforms, and operational events.

The Blockchain system combines the IoT-enabled services assuming integrity and has productivity as per the demand of supply chain exchange partners. The combinations of both technology help to prepare guidelines to work in a more responsible and distributed peer-to-peer system in a secure and real-time environment [10]. The Blockchain supports various challenges implicit in IoT device networks i.e. Unique ID, trust management, information tracking accountability, and access controls. The integrated IoT solution with Blockchain could help customers to reject the purchase of forged ownership of products.The IoT security model is as shown in Figure 9.13.

9.3.2 Scalability

The sensors in the IoT node have limited computing power and expansion to upgrade for a scalable issue. The solutions can be possible using Blockchain to address the

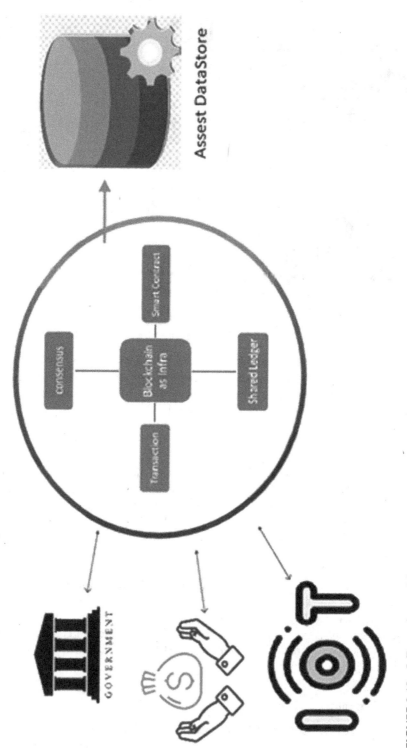

FIGURE 9.13 IoT security model.

Blockchain and IoT Optimization

scalability requirement of IoT in the supply chain. The consensus mechanism and structure are compatible with IoT applications up to a certain extent. Therefore private or consortium Blockchain can be viewed as highly beneficial for supply chain application due to a limited number of nodes, which can be applied for IoT data filtering o increase scalability after Blockchain. The applications can be useful for IoT networks for smart contracts increasing scalability and transaction per second.

Fragmentation is a novel approach to facilitate scalability by distributing the contents of the information to all the nodes in the network. Blockchain partitioning for IoT enables supply chain like the main chain for global events i.e. transshipments, disaster monitoring containing operation and energy plan while secondary chain for local events transactions i.e. inventory control, inbound logistics, production monitoring, etc for frequent local transaction record. The scalability model is as shown in Figure 9.14.

9.3.3 Immutability and Auditing

The support chain automation can be advanced by integrity Blockchain technology with IoT devices which creates an immutable transaction ecosystem for improving audits. The smart IoT data continuously pose autonomously create an ecosystem consisting of immutable transition sectors useful for product traceability and authentication process. Since sensors value adds additional trust incorporating real-time and immutable data, Blockchain technology assures micro audits.

As the supply chain experience rapid IoT device usage, editing Blockchain may result in an error in the chain block. As this transaction replication across Blockchain network could be edited or removed immutable data. The protection for products includes users' data for the products which are already dispatched, consumed, deactivated sensors, reusable sensor tags, or outdated technology using Blockchain. The companies can check possibilities for mutable Blockchain, authorizing partners to modify past blocks for editing hash pointers linked in a block. IoT with Blockchain for Agricultural Supply Chain is as shown in Figure 9.15.

9.3.4 Effectiveness and Efficiency

Blockchain helps for smart tracking of physical assets and good in multiplying the supply chain. The authorized partner exchanges information related to assets, products, and status whether in-transit or in-store. The physical asset movement tracking information helps a firm to better control its supply chain ecosystem. The firms are marketing their products to relevant customers through online or mobile devices. The information related to products may be brand, allergens ingredients, processing methods, and claim policy with Blockchain technology, firms assures transaction, authenticity, and transfer of ownership of the products.

The combination of IoT and Blockchain helps to provide a reliable environment for improving legal agreements, securing the identity of IoT devices, etc. The confidential data regarding plant types of equipment, failure predictions, proactive maintenance strategies, shared through IoT devices can be protected through Blockchain. The universal digital ledger is as shown in Figure 9.16.

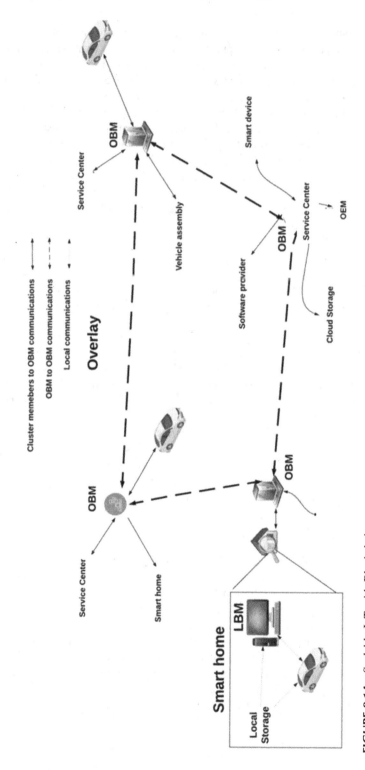

FIGURE 9.14 Scalable IoT with Blockchain.

Blockchain and IoT Optimization

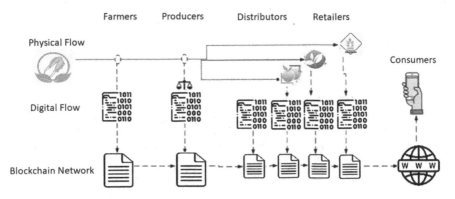

FIGURE 9.15 IoT with Blockchain for agricultural supply chain.

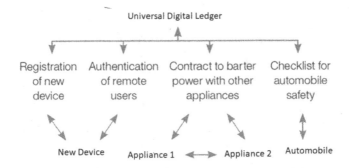

FIGURE 9.16 Universal digital ledger with IoT.

The supporting service providers may suppliers of the spare parts, regulation, service providers may inspect and certify the availability and utilization based on authorized information through Blockchain. Hence Blockchain mechanism with IoT enables business is acting as a catalyst for increasing machine diagnosis, data analysis, and overall maintenance process.

9.3.5 TRACEABILITY AND INTEROPERABILITY

The tamper-evident RFID tags are now day's used with Blockchain technology for verification of the source and authenticity of the product. The tag fixed on items may insecure and find refilling of contents with duplicate products. The authenticity of the product can be cross-checked by logging data on the Blockchain. The history and original production location can be traced using product ID in the system. The absolute information about eatables or drinking products can be confirmed only in physical testing laboratories instead of scanning original authenticity on the outer package label information. The product traceability and interoperability is as shown in Figure 9.17.

There are many project use combination of Blockchain with IoT for improved traceability and interoperability. For example foods, consumer goods, pharmaceutical sectors, etc. A company like SKYCELL developing a system blended with

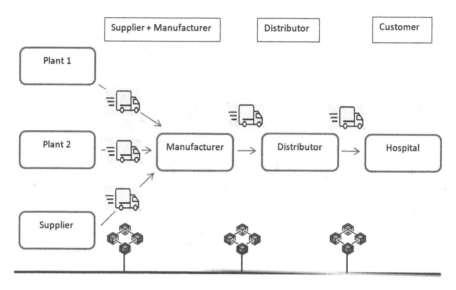

FIGURE 9.17 Product traceability and interoperability.

Blockchain with IoT to monitor temperature, humidity, location, maintaining temperature in their container for refrigerated biopharmaceuticals. The IoT network system with Blockchain enables secure & frictionless dialog between distributed ledger and sensors for quality assurance & supply chain visibility optimizations.

9.3.6 Quality of Service

The traditional system much relay on maintaining consistent data provenience for originality, owner, and transformation of data information. The Blockchain system in IoT networks for the execution of process on IoT technology platform in a decentralized environment for these quality improvements. Through Blockchain, medium metadata is protected from unsecured and unofficial revelation. In a cloud environment, data provenience can be improved record data on a cloud node over the distributed network with a strong fault tolerance ledger.

The Blockchain easily merges big data in support chain use cases with necessary consequences suggestion, supply checking enables digitally-led and process-centric. Big data is facing three genuine challenges i.e. control, data authority, and data monetization. Blockchain technology supports control to govern multi-party structure, reliability, trustworthiness, and ownership of the data transformation process in the universal data marketplace. The quality assurance components can provide services, improvements, satisfactions and guarantee as quality of services as shown in Figure 9.18.

9.4 ENERGY OPTIMIZATION AND BLOCKCHAIN MECHANISM

The Blockchain mechanism supports IoT networks for creating a trusted platform for data communications. The Blockchain mechanism process required heavy computing

Blockchain and IoT Optimization

FIGURE 9.18 Quality of services.

for supporting rigid security. The computing process of Blockchain can be possible to optimize by distributing tasks over the edge node of a network. The hash computing task can be allocated to nodes nearby or adjacent to the edge node and encrypted hashes of block can be cashed in the border or edge gateway. The target of heavy computing of hash in a distributed manner at the edge node is to optimize the load at gateways and accordingly save energy in the network.

9.4.1 Optimization Process

The Blockchain mining process is distinguished over an adjacent node and border gateway node and contents of cryptographic hashed information are shared and cashed in the border gateway node. The minor in the blockchain process creates a new block of chain with unique nonce i.e. 32-bit identifier information and hash by referring to the hash 256-bits security information value of the previous block.

The minor process can be distributed over nearby or adjacent network nodes and gateway nodes. The process can be an optimized load of the existing node for the heavy computing process. The criteria to select adjacent or nearby nodes is based on the energy state and storage capacity. By comparing the information, the adjacent node compute to get the permission of minor computing as per the status of availability, priority is computed; accordingly, the job is distributed over adjacent nodes. So while forwarding information it checks with cache stored hash value and authentication of flows.

9.4.2 Resource Management Using Blockchain

For guaranteed security and privacy in IoT network applications, Blockchain ensures priority choices for companies and consumers. There are some issues with Blockchain computing in IoT networks like more energy is required for heavy computation of minors. The consensus mechanism creates extra computation overhead in the IoT network. The issue can be resolved by improving the efficiency of the consensus process.

The energy consumptions and computational overhead can minimize by various approaches for optimization in IoT networks. It is possible to optimize the consumption of energy by using Markov Decision Process (MDP) approach optimization model. The master control node adjacent to the gateway disturbed the load to adjacent or nearby nodes, blocks size, and edge node gateway is selected for optimizing device energy with reduced system cost.

9.5 ENERGY OPTIMIZATION IN BLOCKCHAIN-ENABLED IoT NETWORKS

The IoT networks efficiently work between machines and humans, humans and humans, machines and machines. It's important to ensure the security and authentication of data in an IoT network. Also, there are certain constraints at the computational and energy resources level. The nodes with minimum battery capacity and limited computing resources make the system hard to compute and complex. Energy optimization in Blockchain-enabled IoT Networks is as hwon in Figure 9.19.

The security issues can be resolved by deploying a backbone mechanism but need heavy computations. The distribution of computing components can be possible to allocate to promising nodes in the network. The efficiency can be improved by these computational distributions of tasks to relatively reduce energy consumption in the device network. As compared to centralized computing at a centralized cloud system, distributed computing system has advantages i.e. lower computational heads, shorter transmission latency, and optimum energy consumptions. Also, a distributive computing environment enables the verification process of Blockchain efficiency. There can be ensured verification process with better efficiency and stability.

The Integration of Blockchain with edge computing is providing a supportive environment to companies and consumers. There are certain challenges and issues for joint considerations of both Blockchain and distributive edge node computing. The strategy balance energy consumption issue and computational overload by selecting a permissible nearby server to control execution task. The energy can be optimized by focusing on the controller part for distribution mechanism, load distribution task decision, block size, and server to be selected with optimal distance. The energy consumption can be optimized by reducing computational overloads in the system.

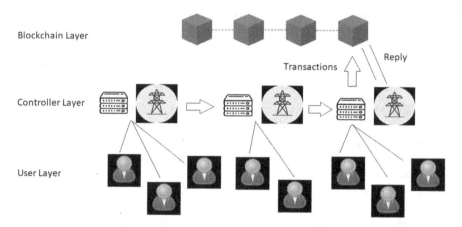

FIGURE 9.19 Energy optimization in Blockchain-enabled IoT networks.

The nodes in the network from a variety of application domains like smart cities, data acquisition systems, monitoring systems that are distributed over local networks. The process can be initiate by allocating computational tasks to local controllers for data consensus and records. The applications in the network can be compartmentalized in terms of cells or units and each cell or Unit consisting of IoT device nodes for a particular application part. From each cell or unit, the edge node controller is called Edge Node Controlling Server (ENCS). The local server for controlling data communication in a unit called Local Node Controlling Server (LNCS), which receives and sends data and processes in itself.

The set of cells with dedicated controllers LNCE and ENCS linked to another cell for data communication. As per the capacity of corresponding nodes in a cell, the task for computing cache is distributed and cryptographic hashed code cached in ENCS of that call. Then each cell's ENCS share block information created to preserve the security of data in the network also as per the capacity of battery storage, information also shared in-network for selection of appropriate node to participate in heavy computation and energy life enhancement in the network.

9.6 SUMMARY

This chapter introduces Blockchain technology and IoT domains use cases. It focuses on Blockchain terminologies, mechanisms, and the distributed P2P networking concept. Blockchain support for IoT applications are described for various use cases. The different characteristics of IoT networks with Blockchain are also described. The processes for energy optimization and resource management using Blockchain are elaborated. Finally, energy optimization in Blockchain-enabled IoT networks is explained.

Exercise
1. Explain Blockchain technology and IoT convergence.
2. Describe Blockchain mechanism with a suitable example.
3. Explain Distributed P2P Networking.
4. How does Blockchain support IoT applications?
5. Explain the various supports of Blockchain concept in IoT applications.
6. What are the different characteristics of IoT networks with Blockchain?
7. Explain the optimization process for energy optimization using Blockchain.
8. Explain resource management using Blockchain.
9. Explain energy optimization in Blockchain-enabled IoT networks.

REFERENCES

[1] Mahamure, S. W. S. 2020. Blockchain technology for real estate documents protection using ethereum. *2020/5 Journal of Future Generation Communication and networking*, 13(3):1287–1295 Publisher, ISSN: 2233-7857 IJFGCN SERSC.

[2] Dinesh Kumar, K., Senthil, P. and Manoj Kumar, D. S. march 2020. Educational certificate verification system using Blockchain. *International Journal of Scientific and Technology research*, 09(03), ISSN 2277-8616 82 IJSTR©2020.

[3] https://dev.to/grantwatsondev/understanding-blockchain-technology-56n7
[4] Lee, G. M. *A Blockchain-based Trust System for Decentralised Applications: When trustless needs trust*. Future Generation Computer Systems: the international journal of grid computing: theory, methods and applications. ISSN 0167-739X (Accepted)
[5] https://builtin.com/blockchain
[6] https://www.csusm.edu/el/programs/science-technology/certblockchain/index.html
[7] Häubl, Christian, Wimmer, Christian, and Mössenböck, Hanspeter. 2012. "Evaluation of trace inlining heuristics for Java". In *Proceedings of the 27th Annual ACM Symposium on Applied Computing (SAC '12)*. Association for Computing Machinery, New York, NY, USA, pp. 1871–1876.
[8] https://www.swinburne.edu.au/study/courses/units/Blockchain-Business-Models-and-Applications--INF80050/local
[9] Zhang, R., Xue, R. and Liu, L. 16 Aug 2019. "Security and Privacy on Blockchain" arXiv:1903.07602v2 [cs.CR]
[10] Rejeb, A., Keogh, J. G. and Treiblmaier, H. 2019. Leveraging the Internet of Things and Blockchain technology in supply chain management. *Future Internet*, 11:161.

Index

Note: *Italicized* and **bold** page numbers refer to figures and tables.

ABC *see* Artificial Bees Colony
ACO *see* Ant Colony Optimization
active attacks, 7
Adaptive Self-Configuring Sensor Network
  Topologies (ASCENT), 29–31, *30*,
  46–47
Ad Hoc on Demand Vector (AODV), 56
AEH *see* ambient energy harvesting
AIS *see* Artificial Immune System
AJIA (adaptive joint protocol based on implicit
  ACK), 133
ambient energy harvesting (AEH), 3
ANNs *see* artificial neural networks
anomaly detection, 134
Ant Colony Optimization (ACO), 181, *196*, *198*
antigens, 200
AODV *see* Ad Hoc on Demand Vector
application layer, 11, 84–85
application level optimization, 170–171, *171*
Artificial Bees Colony (ABC), 197
Artificial Immune System (AIS), 199–201, *203*
artificial neural networks (ANNs), 193, *194*, *195*
ASCENT *see* Adaptive Self-Configuring Sensor
  Network Topologies

Bee Colonies, 181
Bellman-Ford shortest path algorithm, 23
bio-inspired computing, 181–203, *182*;
  approaches, for optimizations,
  188–203, **190–191**; Artificial Immune
  System, 199–201, *200*; artificial neural
  networks, 193, *194*, *195*; epidemic
  spreading, 201–203, *203*; evolutionary
  algorithms, 189, 191–192, *192*; firefly
  algorithm, 198–199, *199*; flexible
  infrastructure, 186; intelligence
  analysis, 186; motivation for, 184–188;
  self-adaptation, 185–186, *187*; self-
  healing ability, 186–188, *188*; self-
  organization, 184–185, *185*; software
  control automation, 186; Swarm
  intelligence, 196–197, *197*
bio-inspired engineering, 183–184, *183*
bio-inspired intelligent algorithms, *189*
bio-inspired process, *182*, 183
BLE *see* Bluetooth low energy protocol
Blockchain, 205–223, *206*; auditing, 217, *217*;
  distributed peer-to-peer networking,
208, 208–209; effectiveness, 217, 219,
  219; efficiency, 217, 219, *219*;
  immutability, 217, *219*; In-car payment
  model, 212–213, *213*; interoperability,
  219–220, *220*; with IoT characteristics,
  214–220, *215*; manufacturing
  maintenance support, 210, *211*;
  mechanism, 206–208, *207*;
  mechanism, energy optimization and,
  220–222; quality of service, 220, *221*;
  resource management using, 221–222;
  scalability, 215, 217, *218*; securing IoT
  networks, 209–210; security, 215, *216*;
  self-sovereign identity, identity
  authentication using, 213–214, *214*;
  structure of blocks, *206*; support IoT
  applications, 209–214; traceability,
  219–220, *220*; transparency in supply
  chain, 212, *212*; vehicle insurance
  model, 213, *214*
block neighborhood, 138, *139*
Bluetooth, 5
Bluetooth low energy (BLE) protocol, 168
BOUNDHOLE, 71

CARP *see* channel-aware routing protocol
CBTC *see* Colored Based Topology Control
  algorithm
CCA *see* cyclic cellular automata
CCP *see* coverage configuration protocol
CDS *see* connected dominating set
cellular automata-based topology control
  algorithms, 54, 137–153; cyclic
  cellular automata, 143–152; for IoT
  application, 152–153; sensor network
  clustering, 140; sensor networks, 139;
  weighted Margolus neighborhood,
  140–141; weighted Moore
  neighborhood, 141–143, *142*, *143*
channel-aware routing protocol (CARP), 13,
  60, 64
cloud computing, 84
clustering mechanisms, 125
clustering topology control approach, 31–36,
  47–49, **52**
CLUSTERPOW, *39*, 39–40, 50, 51
CoA *see* coefficient of variance approach
CoAP *see* constrained application protocol

225

Index

CoCoA *see* congestion control advance approach
coefficient of variance approach (CoA), 174
Cognitive RPL (CORPL), 12, 59–60, 64
collection tree protocol (CTP), 13
Colored Based Topology Control (CBTC) algorithm, 73
communication architecture, 4–7
COMPOW, 25–26, 39, 40, 44, 50, 75
computational heterogeneity, 55
congestion control, 174, *175*
congestion control advance (CoCoA) approach, 174
connected dominating set (CDS), 31; energy-efficient distributed, 32–35, 48, 49; energy-efficient topology control algorithms, 121–125, *123*, 126–129, *129*; power aware, 31–33, *32*, 47–49, 51
constrained application protocol (CoAP), 164, 168, 171
context-aware routing in, IOT networks, 60–63, **62–63**
CORPL *see* Cognitive RPL
coverage configuration protocol (CCP), 73
CPU, 3
critical transmission range (CTR) problem, 74, 76–77
CTP *see* collection tree protocol
CTR *see* critical transmission range problem
cyclic cellular automata (CCA), 143–152; -based topology control algorithm, 144–146, *146*; data aggregation model, 150; entropy base model, 150–152, *151*; Greenberg-Hastings model, 147; mathematical analysis, 147–150; mathematical model, 146–147; proposed, 147
cyclic self-reproduction system, 137, 143, 151

DAO *see* Destination Advertising Object/Destination Advertisement Object
DAO Acknowledgment (DAO-ACK), 56, 57
DAO-ACK *see* DAO Acknowledgment
DARPA *see* United States Defense Advanced Research Projects Agency
data aggregation model, 150, 162–163, *162*
data gathering, 6
Defense applications, 9
Destination Advertising Object/Destination Advertisement Object (DAO), 56–59
Destination-Oriented Directed Acyclic Graph (DODAG), 55–56; with non-sorting mode, *58*, 59; with sorting mode, *58*, 59
device level optimization, *167*, 167–168
device reliability, 130

DIO *see* DODAG Information Object
DIS *see* DODAG Information Solicitation
Distributed Ledger Technology (DLT) *see* Blockchain technology
distributed peer-to-peer networking, *208*, 208–209
Distributed Sensor Network (DSN), 1–2
DLT *see* Distributed Ledger Technology
DODAG *see* Destination-Oriented Directed Acyclic Graph
DODAG Information Object (DIO), 57, 59
DODAG Information Request, 57
DODAG Information Solicitation (DIS), 56, 57
dominant set (DS), 31
downward routing, 58–59
DRX (discontinues reception), 177
DS *see* dominant set
DSN *see* Distributed Sensor Network
DSS (distributed self-spreading) algorithm, 74
DTX (discontinues transmission), 177
DVFS *see* dynamic voltage frequency scaling
dynamic voltage frequency scaling (DVFS), 117

EAP *see* extensible authentication protocol
EAs *see* evolutionary algorithms
E-CARP, 14, 60
ECDS *see* energy-efficient distributed connecting dominating sets
ED *see* Event Detection
Edge Node Controlling Server (ENCS), 223
ENCS *see* Edge Node Controlling Server
energy conservation, 172–174, *173*
energy consumption, 6, 14
energy consumption in IoT networks, challenges for, 14–16; árchitecture design, 16; IoT standards, 16; safety challenge, 15–16; subsystems, 15; user privacy, 15
energy consumption model, 116
energy-efficient distributed connecting dominating sets (ECDS), 32–35, 48, 49
energy-efficient scheduling algorithms, 110–111, *114*, 114–118, 116–118; global algorithm, 117; local algorithm, 117–118, *118*
energy-efficient topology control algorithms, 121–134; clustering mechanisms, 125; connected dominating set, 121–125, *123*, 126–129, *129*; implementations, 129–130, *131*, *132*; network model, 125–126
energy heterogeneity, 55
energy model, 78–79
energy optimization: in Blockchain-enabled IoT networks, 222, 222–223; and blockchain mechanism, 220–222
epidemic spreading (ES), 201–203, *203*
ES *see* epidemic spreading

Index

Event Detection (ED), 6
evolutionary algorithms (EAs), 189, 191–192, *192*
extensible authentication protocol (EAP), 175

FA *see* firefly algorithm
failure management, 170
fault tolerance, 5–6, 155, *156*
FETC, 99–110, *103–111*, **112**, **113**, *114*
FETCD, 99–110, *103–111*, **112**, **113**, *114*
firefly algorithm (FA), 198–199
flexible infrastructure, 186

Gabriel graph (GG), 71, 99, *102*
GAF *see* Geographical Adaptive Fidelity
gateway placement, 110–111, *114*, 114–118
gateways, 4
GEAR *see* geographical and energy aware routing
generalized Petri stochastic net (GSPN), 133
Geographical Adaptive Fidelity (GAF), 26–28, *27*, *28*, 45, 47
geographical and energy aware routing (GEAR), 71
GG *see* Gabriel graph
GHM *see* Greenberg-Hastings model
global algorithm, 117
Global Positioning System (GPS), 42
GPS *see* Global Positioning System
GPSR *see* greedy perimeter stateless routing
graph, definition of, 95
graphic routing, 71; schemes, **72**
greedy perimeter stateless routing (GPSR), 71
Greenberg-Hastings model (GHM), 147
GSPN *see* generalized Petri stochastic net

hardware interface layer, 11
healthcare applications, 9
HEED, 51
heterogeneity, handling, 157, *158*
heterogeneity of network technologies (HetIoT), 82–88, *83*; application layer, 84–85; cloud computing, 84; intelligent transportation system, 87; network layer, 84; sensing layer, 83–84; smart agricultural, 86; smart healthcare, 87; smart home, 86–87; smart industrial, 85–86
heterogeneous operations, administration and management (H-OAM) approach, 174–175, *176*
heterogeneous topology control (HTC) algorithm, 54–55
Heterogeneous Wireless Sensor Network (HWSN), 54–55
HetIoT *see* heterogeneity of network technologies
H-OAM *see* heterogeneous operations, administration and management approach
holes problems, 71–73
homogeneous *vs.* heterogeneous networks, 6–7
homogenous network model, 76–77, *78*
hop model, 78
HTC *see* heterogeneous topology control algorithm
HWSN *see* Heterogeneous Wireless Sensor Network
hybrid star-mesh network topology, 20, *21*
hybrid topology control approach, 36, 38–39, *38*, **52**; evaluations based on network lifetime definitions, 51

IDSs *see* Intrusion Detection Systems
IEEE, 2
IEEE 802.15.4, 2, 5, 14
IETF, 12, 64
In-car payment model, 212–213, *213*
industrial applications, 9
Industry 4.0, 85–86
INF *see* intermediate node forwarding
infrastructure to infrastructure (I2I), 8
intelligence analysis, 186
intelligent transportation system (ITS), 87
intermediate node forwarding (INF), 71
Intermediate System to Intermediate System (IS-IS), 56
Internet of things (IoT), 1; applications of, 8–9; bio-inspired computing, 181–203; Blockchain, 205–223; context-aware routing in, 60–63, **62–63**; domain, simulation using MATLAB for, 81–82; energy consumption in, challenges for, 14–16; layer stack, 9–11; performance optimization, 155–179; protocols, 11–14; protocol stack layers, *10*; reliability, 130–134; and routing protocols, 55–60; topology control methods, 21–55
Internet of vehicles (IoV), 8
intra-vehicle applications, 9
Intrusion Detection Systems (IDSs), 7
IoT *see* Internet of things (IoT)
IoV *see* Internet of vehicles
IPOLY algorithm, 127, 129
IPv6 routing protocol, 14, 164, 172, 174
IS-IS *see* Intermediate System to Intermediate System
ITS *see* intelligent transportation system
I2I *see* infrastructure to infrastructure

JACKPAC, 81
JAM protocol, 72

LBR *see* LLN boundary router
LEACH *see* Low-Energy Adaptive Clustering Hierarchy
LEBTC *see* link efficiency-based topology control
lifetime and latency aggregatable metric (L2AM), 177
Lightweight On-Demand Ad Hoc Distance Vector Routing – Next Generation (LOADng), 12–13
link efficiency-based topology control (LEBTC), 53, *101*, 104, *113*; algorithm, for IoT domain application, 93–119; energy consumption model, 116; energy-efficiency scheduling algorithm, 110–111, *114*, 114–118, 116–118, *118*; flow diagram, 99, *100*, *101*; gateway placement, 110–111, *114*, 114–118; implications of, 99–110, **101**, *102*, *103*; improved algorithm, 96–99; mathematical model, 98–99; network model, 94–96, **96**; placement of gateways, *115*, 115–116; received signal strength indicator, 93–94; received signal strength indicator, limitation of, 94; task model, 116
link heterogeneity, 55
link quality indicator (LQI), 133
LLN boundary router (LBR), 58, 59
LLNs *see* Low power and Lossy Networks
LNCS *see* Local Node Controlling Server
LNSM *see* log-normal shadowing model
load balancing, 170
LOADng *see* Lightweight On-Demand Ad Hoc Distance Vector Routing – Next Generation
local algorithm, 117–118, *118*
Local Node Controlling Server (LNCS), 223
log-normal shadowing model (LNSM), 94, 100
long-distance path model, 77–78
Low-Energy Adaptive Clustering Hierarchy (LEACH), 40–41, 43, 50–51
Low power and Lossy Networks (LLNs), 56–59
LQI *see* link quality indicator
L2AM *see* lifetime and latency aggregatable metric (L2AM)

MAC protocol, 77
Margolus neighborhood, 131–132, *132*
Markov Decision Process (MDP), 221
MATLAB for IoT domain, simulation using, 81–82
maximum independent set (MIS), 32–34, 48, 121
MDP *see* Markov Decision Process
MDS *see* minimal dominating set
mean time between failures (MTBF), 130
mean time to failure (MTTF), 130

mean time to repair (MTTR), 130
MECN *see* Minimum Energy Communication Network
MEMS *see* Micro-Electrical Systems
mesh network topology, 20, *21*
message queuing telemetry transport (MQTT), 168, 170–171
microcontroller, 3
Micro-Electrical Systems (MEMS), 7
minimal dominating set (MDS), 48
minimum distance tree (MST), 77
Minimum Energy Communication Network (MECN), 23, 24, 41–45
MIS *see* maximum independent set
mobile network, 74
Moore neighborhood, 137, *138*
MP2P *see* multipoint to point
MQTT *see* message queuing telemetry transport
MQTT-SN protocol, 171
MST *see* minimum distance tree
MTBF *see* vmean time between failures
MTTF *see* mean time to failure
MTTR *see* mean time to repair
multipoint to point (MP2P), 56

National Science Foundation (NSF), 81
neighbor, definition of, 95
network coverage, 164–165, *165*
network layer, 84
network level optimization, 168–170, *169*
network models: energy-efficient topology control algorithms, 125–126; energy model, 78–79; homogenous model, 76–77, *78*; hop model, 78; long-distance path model, 77–78; wireless propagation model, 77
network rehabilitees, definition of, 95–96
network reliability, 133, 176–177, *177*
network routing, 172
network topology, 14
network visibility, 160, *161*
NSF *see* National Science Foundation
NS2, 80–81

OGDC *see* optimal geographical density control
OLSR *see* Optimized Link State Routing
OMNEST, 80
OMNeT++, 79–80
Open Shortest Route First (OSPF), 56
operating systems (OS), 4
operational conditions monitoring, 9
optimal geographical density control (OGDC), 73
Optimized Link State Routing (OLSR), 56
OS *see* operating systems
OSI (open system interconnection) model, 9–11; application layer, 11; hardware

Index

interface layer, 11; physical or sensor layer, 10; processing and control layer, 10; RF layer, 11; session/message layer, 11; user experience layer, 11
OSPF *see* Open Shortest Route First

PACDS *see* power aware connected dominating set
Packet error rate (PER), 60
Particle Swarm Optimization (PSO), 196
passive attacks, 7
pathogens, 199
PCs *see* personal computers
PDAs *see* personal digital assistants
PEAS protocol, 73
PER *see* Packet error rate
performance optimization, 155–179; application level optimization, *170*, 170–171; congestion control, 174, *175*; data aggregation, *162*, 162–163; device level optimization, *167*, 167–168; energy conservation, 172–174, *173*; fault tolerance, 155, *156*; heterogeneity, handling, 157, *158*; levels of, 166–171; network coverage, 164–165, *165*; network issues, 155–160; network level optimization, 168–170, *169*; network reliability, 176–177, *177*; network routing, 172; network visibility, 160, *161*; quality of service, 177–178, *178*; restricted access, 160; routings in IoT networks, *163*, 163–164; scalability, 175, *176*; security enforcement, 156–157, *157*; self-configuration, 158, *159*; sensor localization, 165–166, *166*; solutions for, 171–178, *172*; unintended interference, 158–160, *160*
personal computers (PCs), 4
personal digital assistants (PDAs), 4
physical or sensor layer, 10
placement of gateways, *115*, 115–116
point to multipoint (P2MP), 56
point to point (P2P), 56
power adjustment approach, 23–26, *24*, 41–45, **52**; evaluations based on network lifetime definitions: 44–47
power aware connected dominating set (PACDS), 31–33, *32*, 47–49, 51
power control mechanisms, 74–76
power management mechanisms, 76, **77**
power mode approach, 26–31, **52**; evaluations based on network lifetime definitions: 49–51
power source, 3
processing and control layer, 10

prolong network lifetime, 170
proof-of-work, 207
PSO *see* Particle Swarm Optimization
P2MP *see* point to multipoint
P2P *see* point to point

QoA *see* quality of information
QoS *see* quality of service
quality of information (QoA), 174
quality of service (QoS), 5, 133, 153, 164, 170, 174, 177–178, *178*; Blockchain, 220, *221*

RA *see* range assignment problem
radio frequency (RF), 4
range assignment (RA) problem, 75
received signal power, definition of, 95
received signal strength indicator (RSSI), 53, 93–94, 99, 104, *107*, *110*, *111*; limitation of, 94
relative neighbor graph (RNG), 99, *102*
reliability of IoT: anomaly detection, 134; device reliability, 130; network reliability, 133; system reliability, 133
RERR *see* route error
RERUM, 177
RESTfull HTTP, 171
restricted access, 160
retransmission time out (RTO), 174
RF layer, 11
RNG *see* relative neighbor graph
ROLL (Routing Over Low-Power and Lossy Networks), 56
round trip time (RTT), 174
route error (RERR), 13
route reply (RREP), 13
route request (RREQ), 13
routing protocol for low-power and lossy networks (RPL), 12, 55–56, 55–59, 164; cognitive, 12, 59–60; definition of, 56; downward routing, 58–59; messages, 57; network topology, 57, *57*; routing with, 57–59, *58*; upward routing, 58
routing protocols: IoT and routing protocols
routings, in IoT networks, *163*, 163–164
RPL *see* routing protocol for low-power and lossy networks
RREP *see* route reply
RREQ *see* route request
RSSI *see* received signal strength indicator
RTO *see* retransmission time out
RTT *see* round trip time

SASA Topology Maintenance Protocol, 71, 72
SAX *see* symbolic aggregate approximation

security: Blockchain, 215, *216*; issues, 7
SEIR (Susceptible, Exposed, Infectious, Recovered) model, 202
self-adaptation, bio-inspired, 185–186, *187*
self-configuration, 158, *159*
self-healing ability, bio-inspired, 186–188, *188*
self-organization, bio-inspired, 184–185, *185*
self-organizing things (SoT), 174
self-sovereign identity (SSI): identity authentication using, 213–21, *214*
sensing layer, 83–84
sensor connectivity topology: power control mechanisms, 74–76; power management mechanisms, 76, **77**
sensor coverage topology: hybrid network, 74; mobile network, 74; multiple coverage, 73–74; partial coverage, 73; single coverage, 73; static network, 73–74
sensor localization, 165–166, *166*
sensor networks: cellular automata for, 139; clustering, 140; topologies *see* sensor network topologies
sensor network topologies: hybrid star-mesh network topology, 20, *21*; mesh network topology, 20, *21*; star network topology (single point-to-multipoint), 19, *20*
sensor node, 2
sensor transducer, 3–4
session/message layer, 11
SI *see* Swarm intelligence
simulation platforms: NS2, 80–81; OMNeT++, 79–80
6LoWPAN, 5
SKYCELL, 219–220
slider neighborhood, 138, *139*
Small Minimum Energy Communication Network (SMECN), 25, 44
smart agricultural, 9, 86
smart cities, 8
smart environmental, 8
smart healthcare, 87
smart home, 86–87
smart industrial, 85–86
SMECN *see* Small Minimum Energy Communication Network
software control automation, 186
SOSUS *see* Sound Surveillance System
SoT *see* self-organizing things
Sound Surveillance System (SOSUS), 1
SPAN, *38*, 38–39, 49–51
Sparse Topology and Energy Management (STEM), 28–29, *29*, 45–46
Spatial Process Estimation (SPE), 6
SPE *see* Spatial Process Estimation
SRA *see* symmetric range allocation

SSI *see* self-sovereign identity
star network topology (single point-to-multipoint), 19, *20*
STEM *see* Sparse Topology and Energy Management
Swarm intelligence (SI), 196–197, *197*
symbolic aggregate approximation (SAX), 175
symmetric range allocation (SRA), 76
system reliability, 133

Task Managers, 4
task model, 116
TDES process, 202
time of data delivery, 6
Tiny OS, 4, 13
TMPO *see* Topology Management by Priority Ordering
TOME method, 87
topology control design problems, 69–76; awareness problem, 71–73; data processing, 69; energy conservation, 69; limited bandwidth, 69; low-quality communications, 69; operating in hostile environments, 69; scalability, 69; sensor connectivity topology, 74–76; sensor coverage topology, 73–74, **75**; taxonomy, 70, *70*; unstructured and time-varying network topology, 69
topology control methods: cellular automata-based topology control, 54; classification, 22–23, *23*; clustering approach, 31–36; comparative analysis, 41–53, **42**, **52–53**; design problems *see* topology control design problems; heterogeneous topology control algorithm, 54–55; hybrid approach, 36, *38*, 38–39; improved reliable and energy efficient, 54; IoT and, 21–55, 53–55; link efficiency-based topology control, 53; power adjustment approach, 23–26, *24*; power mode approach, 26–31; purpose of, 22
Topology Management by Priority Ordering (TMPO), 34, 36, 37, 48–49
transceiver, 4

UDG *see* unit disk graph
Underground Wireless Sensor Network (USWN), 60
unintended interference, 158–160, *160*
unit disk graph (UDG), 122
United States Defense Advanced Research Projects Agency (DARPA), 1, 80–81
upward routing, 58

Index

user experience layer, 11
US Military: Sound Surveillance System (SOSUS), 1
USWN *see* Underground Wireless Sensor Network

VANETs *see* vehicle ad-hoc networks emerge
variable back-off factor (VBF), 174
VBF *see* variable back-off factor
vehicle ad-hoc networks (VANETs) emerge, 87
vehicle insurance model, 213, *214*
vehicles to infrastructure (V2I), 8
vehicles to vehicles (V2V), 8
VINT *see* Virtual Inter Network Test Bed
Virtual Inter Network Test Bed (VINT), 80–81
Von Neumann neighborhood, 137, *138*
V2I *see* vehicles to infrastructure
V2V *see* vehicles to vehicles

weakly symmetric range assignment (WSRA), 76
weighted Margolus neighborhood, cellular automata, 140–141
weighted Moore neighborhood, cellular automata, 141–143, *142*, *143*
wireless propagation model, 77
wireless sensor networks (WSN), 1–7, *2*; applications of, 8–9; communication architecture, 4–7; communication standards and specifications, 5; components, *3*; design factors and requirements, 5; functionality of, 2; gateways, 4; protocols, 11–14; security issues, 7; Task Managers, 4; taxonomy of topology issues in, 70, *70*; technology, 2–4
WSN *see* wireless sensor networks
WSRA *see* weakly symmetric range assignment

XMPP, 171

ZigBee, 5, 20, 174
ZigBee Alliance, 2